A Voyage to Newfoundland

A Voyage
to Newfoundland

Julien Thoulet

Translated from the French and edited by
SCOTT JAMIESON

McGILL-QUEEN'S UNIVERSITY PRESS   Montreal & Kingston • London • Ithaca

© McGill-Queen's University Press 2005

ISBN 0-7735-2867-9

Legal deposit second quarter 2005
Bibliothèque nationale du Québec

Printed in Canada on acid-free paper that is 100% ancient forest
free (100% post-consumer recycled), processed chlorine free.

This book has been published with the help of a grant from the
Canadian Federation for the Humanities and Social Sciences,
through the Aid to Scholarly Publications Programme, using funds
provided by the Social Sciences and Humanities Research Council
of Canada.

McGill-Queen's University Press acknowledges the support of the
Canada Council for the Arts for our publishing program. We also
acknowledge the financial support of the Government of Canada
through the Book Publishing Industry Development Program
(BPIDP) for our publishing activities.

Library and Archives Canada Cataloguing in Publication

Thoulet, J. (Julien), 1843–1936.
A voyage to Newfoundland / translated from the French and edited
by Scott Jamieson.

Includes bibliographical references and index.

ISBN 0-7735-2867-9

1. Thoulet, J. (Julien), 1843–1936—Travel—Newfoundland and
Labrador. 2. Clorinde (Ship) 3. Newfoundland and Labrador—Description and travel. 4. Newfoundland and Labrador—History—19th
century. 5. Cod-fisheries—Newfoundland and Labrador—History—
19th century. 6. Saint Pierre and Miquelon—Description and travel.
7. Cape Breton Island (N.S.)—Description and travel. I. Jamieson,
Scott, 1952– II. Title.

FC2167.3.T48 2005   917.1804'2   C2004-905441-4

Designed and typeset by oneonone in 10.5/12.5 New Baskerville

For Anne and Elise

Contents

Illustrations

Photographs, except those on pages xiv and 10, are by
  Julien Thoulet
Illustrations, except as noted, are © Société de géographie,
  184 boulevard Saint-Germain, 75006 Paris

Acknowledgments

For some of the surviving generation of "golden agers" who lived through the early and mid twentieth century in France, the name Terre-Neuve evokes the *grande pêche*, when hundreds of ships and thousands of men left the fishing villages of Normandy and Brittany to fish cod on the Grand Banks off Newfoundland, a carryover from the French presence here, which of course is centuries old. My interest in this connection grew out of the realization that the idea of Newfoundland – Terre-Neuve – is still alive to some degree in France and that the historic ties are abundantly documented in French.

Before the new Bibliothèque de France opened at Tolbiac a few years ago, I worked for a time in the old Bibliothèque nationale on rue de Rivoli in Paris. Downstairs in the cool *sous-sol* was the catalogue room, which housed countless bibliographies, inventories, lists, and repertories – essentially the table of contents of France's entire public collection of manuscripts, books, and printed materials of all sorts. My work at the time brought me back there daily for many weeks. In the afternoons, when my concentration began to wane and I needed to get up and move around in the cavernous yet silent room, I would browse through bookshelves, filing cabinets, tables, and the like. I happened on a few titles containing Terre-Neuve, books published in French in the 1800s and earlier. I was soon reading and translating into English a number of first-hand accounts of shipwrecks in the waters off Newfoundland. I found it exciting to be seeking out texts that would be worth sharing with English readers through translation, and I found as many as I could.

I am deeply grateful to Ron Rompkey of Memorial University's English Department for helping me get started on the Thoulet translation, for his enthusiasm and encouragement, and for his expertise every time I needed it. Thanks to him, I also had the privilege of entering into correspondence with Jacqueline Carpine-Lancre, formerly of the Musée Océanographique de Monaco, who generously shared her knowledge of Thoulet, his life and times, and his contributions to French oceanography. To Jacqueline, I owe a huge debt of gratitude. She pointed out many errors in my work; she also guided me repeatedly toward little-known and hard-to-find manuscripts, letters, documents, and articles by or relating to Thoulet. As a specialist in the history of oceanography and a librarian of vast experience, she offered guidance to the catalogues and treasures of the National Archives. From the outset, her unfailing encouragement has been invaluable.

In Corner Brook, Olaf Janzen gave openly of his expert knowledge of French Shore history. He read the manuscript, closely, identified supplementary reading sources, and pointed me toward important background information that I would not likely have found on my own.

Others were generous in their guidance and diligent in suggesting improvements to the manuscript, in detecting flaws and omissions, and in answering my many queries. In particular, I acknowledge these people: my wife, Anne Thareau; Martin Ware, Rainer Baehre, Michael Coyne, David Freeman, Bill Iams, Nicholas Novakowsi, Michael Parker, Holly Pike, Geoff Rayner-Canham, Ron Richards, and David Morrish at Grenfell College; Tony Chadwick, Michael Wilkshire, Gordon Hancock, Philip Hiscock, and James Hiller at Memorial University; Joan Ritcey and Sue-Anne Reid, at the Centre for Newfoundland Studies, Queen Elizabeth II Library, and Dianne Taylor, Memorial University of Newfoundland; Elizabeth Behrens and Louise Gillis at the Ferris Hodgett Library, Sir Wilfred Grenfell College; Carl Best at the Newfoundland and Labrador Provincial Archives; Olivier Loiseaux at the Bibliothèque Nationale de France; Pauline Hayes and Pam Parsons at Sir Wilfred Grenfell College for their research, technical, and secretarial help and support; Alan Farrell at Arts Computing; Charles Conway and David Mercer at the MUN Cartography Lab; and Patricia Churchill and Paula Alexander of the Department of French and Spanish, Memorial University of Newfoundland.

A special word of thanks to Matthew Kudelka for editing the manuscript and to Joan McGilvray and Aurèle Parisien at McGill-Queen's.

I acknowledge the Penguin Group (UK) for use of James Michie's translation of "The Oyster and the Litigants."

Funding was provided by the Canadian Federation for the Humanities and Social Sciences (Aid to Scholarly Publications Programme), the J.R. Smallwood Foundation, Sir Wilfred Grenfell College, and Memorial University of Newfoundland.

SJ
Department of French and Spanish,
Memorial University of Newfoundland

Portrait of Julien Thoulet

Introduction

ABOUT JULIEN THOULET

Julien Thoulet was born in Algiers in 1843 and after a life devoted to science died in Paris in 1936 at the age of ninety-three.[1] He finished his secondary schooling in Paris, where he received his *baccalauréat ès sciences* and became interested in mineralogy and cartography. He worked in the United States, where he spent some time with the Northern Pacific Railroad near Lake Superior.[2] He returned to France and later worked as a laboratory assistant at the Collège de France under the mineralogist Charles Sainte-Claire-Deville. In 1867, he was proclaimed a member of the Société de géographie de Paris, to which he made several contributions of a cartographic nature (more specifically, of a gnomonic nature).[3] In 1880, he obtained a doctorate in physical science for a thesis titled *Contribution to the Study of the Physical and Chemical Properties of Microscopic Minerals,*[4] which he presented to the Faculty of Science of the University of Paris. Within two years, he obtained a post as *maître de conférences* – lecturer – at the University of Montpellier. In 1884 he was appointed professor of geology and mineralogy at the University of Nancy.

Thoulet's early experience in mineralogy and mapmaking barely suggests the direction his long academic and scientific career would take. Jacqueline Carpine-Lancre has described Thoulet's career as quite original and indeed paradoxical.[5] This future oceanographer was appointed to the University of Nancy, which was nowhere near the ocean, and worked there until retiring from teaching in 1913.[6] Through his writings over many years, Thoulet tried to instil in his contemporaries a greater understanding of the physical laws of the sea, which he and others had begun referring as "océanographie pure."[7]

Thoulet's six-month journey to Newfoundland in 1886 marked a turning point in his career: he had been trained in geology, mineralogy, and cartography; from then on, he would dedicate himself to the burgeoning field of oceanography. Between 1888 and 1899, he was often invited to give lectures: to the public at the Sorbonne; to naval officers at the Observatoire de Montsouris; and to officers at the École des hautes-études de la Marine in Paris. In 1903 and 1904, he gave a series of lectures in Paris under the patronage of Prince Albert I of Monaco; in 1913 and 1914, he offered another series in Paris sponsored by the Ligue Maritime. Thoulet's bibliography is far too long to include here in its entirety, but it is worth noting that he published some laboratory research[8] prior to his Newfoundland journey that cannot properly be included in his oceanographical work, which began with his voyage on the *Clorinde*. After 1886, as his bibliography indicates, he devoted himself entirely to the study of oceans (and to a lesser degree, lakes). He wrote more than four hundred books and scientific articles, only a few of which were unrelated to his work in oceanography.[9]

The exact circumstances of Thoulet's first encounter with Prince Albert are unknown, but it is most likely that his Newfoundland voyage and the lectures he gave about it, together with the articles he published about it,[10] came to the attention of the oceanographer–prince, and that this led to their initial meeting. Thoulet and Prince Albert would soon be working together – specifically during scientific expeditions in 1901 and 1903 aboard the *Princess Alice II*.[11] Deacon, in her account of a meeting in Wiesbaden in 1903, notes that plans were made there to draw a chart of the ocean depths. This would involve collating the measurements taken by various nations, and Thoulet would play a key role in this effort.[12] Prince Albert agreed to lend his financial backing to this project and, in Paris, a general chart of the ocean depths was begun.

Thoulet recognized the value of bathymetric measuring and had shown his skill at it with his 1899 chart of the Azores. He would carry the heaviest responsibility for much of this ambitious new project.[13] Unfortunately, the final stages of the chart's first edition were carried out not by Thoulet but under Charles Sauerwein (1876–1913), a young officer in the French navy and Prince

Albert's new collaborator, who had been appointed Chef de service de la carte générale des océans.[14] Sauerwein's lack of scientific expertise and disregard of Thoulet's memorandum, combined with careless drawing of the charts, led to sharp criticisms by Emmanuel de Margerie (1862–1953) as the first edition began to come out. As a consequence, a second edition became necessary immediately. Thoulet had already declared that an undertaking of this nature would require constant revisions as new data became available. After many interruptions, including the First World War and the death of Prince Albert in 1922, the final sheets of the first edition were printed in 1931. The second edition was never published in its entirety: the responsibility for a project of this magnitude – one that required international cooperation and meticulous organization – had grown too heavy for one individual, or even one nation. At a conference in London in 1919, it was decided that the task would be taken over by the International Hydrographic Bureau. The scientific and commercial success of this endeavour became a reality only with the fifth edition; the General Bathymetric Chart of the Oceans (GEBCO)[15] as it is now called, is an ongoing international enterprise whose future is assured. In 2003 in Monaco, the GEBCO celebrated its centenary.

By 1904, Thoulet had extended his reputation beyond scientific circles. His next book, *L'Océan, ses lois, ses problèmes*, published by Hachette, was an effort to educate France's readers about oceanography. He also undertook to complete a project begun in 1897 to publish an underwater atlas of the coasts of France: *Atlas bathymétrique et lithographique des Côtes de France*.[16] Applying the latest scientific equipment and methods, he began a complete survey of the coasts of France, starting with the Mediterranean. With the support of the Ministère de la Marine[17] and a total of eleven vessels (including the *Rolland* out of Banyuls and the *Eider* of the Institut océanographique de Monaco), the first part of the survey was completed, and five coloured maps were published covering the coast from the Spanish border to Saintes-Marie-de-la-Mer.

Over the next few years Thoulet published many articles, memoirs, reports, and studies relating to the Mediterranean, the Atlantic, the Indian Ocean, and freshwater lakes in Europe. His efforts continued to be recognized and he received many honours and distinctions. For example, he was appointed to the Conseil supérieur des pêches maritimes in 1911 and was elected to the Academy of Sciences of Lisbon in 1912.[18] However, the years leading up to his retirement from the University of Nancy were not entirely happy. One of his most promising students, whom Thoulet refers to only as Monsieur Sudry, died unexpectedly, and another, A. Chevalier, was forced to abandon oceanography, leaving no one to carry on Thoulet's work. After retiring in 1913, Thoulet moved to Algiers, where he lectured in oceanography at the Université d'Algers. However, the support he required in order to carry out his plan to map the coast of Algeria – support that had been promised to him – was not forthcoming. Six months later, he left Algeria and, since he was retired, he no longer had a laboratory in Nancy to which to return. When the First World War broke out, the Germans advanced rapidly toward Nancy and in September Thoulet was forced to flee the city with his family. They spent at least the latter part of the war at Erquy in Brittany. Thoulet's papers and furniture survived; they were in storage in Paris, where he intended to move in order to be closer to his family.[19] He died in 1936, survived by his four children (three of whom were married) but no grandchildren. The whereabouts of Thoulet's papers, which he had apparently left at the Bibliothèque de l'Institut océanographique de Paris – where he was able to continue working well into his eighties – are unknown.

Thoulet's later years were not terribly happy ones. For all his hard work, all his contributions to oceanography, and all the favourable recommendations he garnered, he received the promotion he had been led to expect only on the eve of his retirement. He stagnated most of his career at the rank of *professeur, troisième classe.*[20] Thoulet undoubtedly was a man of scientific and moral integrity with a strong sense of duty, but he was perhaps politically naive. His lifelong, incessant hard work, his travels, his lecture tours, the conferences he attended, the countless articles he wrote for popular journals (perhaps to compensate for his extremely low university salary), and all his frantic intellectual effort suggests a certain instability, a man tormented by the failure of his ambitions.

THOULET'S INTEREST IN OCEANOGRAPHY

By the early 1880s, when Julien Thoulet joined the University of Nancy, the world's oceans were inspiring intense scientific interest in most European nations. This led to a number of ambitious expeditions. The best remembered were mounted by the United States, Germany, the Scandinavian countries, and Great Britain. Perhaps the most famous of all was Britain's *Challenger* expedition of 1872–76. For various reasons, except for the expeditions of the *Travailleur* (1880, 1881, 1882) and the *Talisman* (1883), France's endeavours in oceanography have been mostly forgotten, although French scientists were not lacking in expertise and enthusiasm.[21] The scientific work done in French universities and at the Muséum d'histoire naturelle[22] focused primarily on marine biology. The widespread use of undersea cables increased the need for knowledge of the ocean floor, and beginning in the mid-nineteenth century, a number of privately funded expeditions were carried out to acquire it.[23] The French navy, through its voyages of discovery, had played a major role in scientific advancement in the eighteenth and early nineteenth centuries,[24] but by Thoulet's time it was busy with its routine activities, and it was not until the 1870s that its earlier tradition of scientific exploration was revived. Besides making the usual soundings and seafloor maps, the navy established a policy of collaborating with scientists, allowing them use of its vessels on their many voyages around the world. This led, for example, to new knowledge about the circulation of ocean currents and to the discovery of an undersea trench in the equatorial Atlantic.

Thanks to the efforts of an impassioned amateur, the Marquis Léopold de Folin, a French oceanographic expedition was finally organized in 1880. On the first of its three cruises, which the government funded, the *Travailleur* carried a team of scientists led by Alphonse Milne-Edwards.[25] These voyages and others in the early 1880s brought back more detailed knowledge of the waters and deep-sea life of the Bay of Biscay, the Mediterranean, and the temperate zone of the Atlantic as far as the Sargasso Sea.

The new science of oceanography inspired the imagination of scientists as well as the public. Several things suggest this: the 1884 exhibition at the Muséum d'histoire naturelle, which presented the results of the *Travailleur* and *Talisman* expeditions; the popular success of Jules Verne's 1869 novel *Vingt mille lieux sous les mers* (Twenty Thousand Leagues under the Sea); and the interest sparked by the results of Prince Albert's four expeditions aboard the schooner *Hirondelle,* which were presented at the Monaco pavilion during the 1889 Paris World's Fair. Even so, ocean science was an expensive task, and largely for that reason it was unable to find any government backers. Carpine-Lancre suggests another reason: the political climate in France in the 1880s. The country was still reeling from the collective shock of its defeat in the Franco-Prussian War (1870–71), and this made it almost impossible, for political, diplomatic, and financial reasons, to excite the government's interest in scientific or technical enterprises of any kind.[26] Thoulet would have his work cut out for him.

The scientific world at large, the French scientific community, and certainly Thoulet himself would have been keenly aware of the results of the *Challenger* expedition, which had just been published,[27] along with those of the *Travailleur* and the *Talisman*. The reports from the *Challenger* may well have been Thoulet's bible.[28] Thoulet was abreast of the latest developments in ocean science. He had been interested since the early 1880s in the sponge samples collected by the *Talisman*, and he had already published an analysis of them.[29] For whatever reasons, oceanography attracted him, and he was already a keen observer in the field. Clearly, he was not going to willingly spend all his time in his laboratory in Nancy. Although his personal resources were modest,[30] he must have already been preparing his own ocean expedition.

SOME BACKGROUND TO THOULET'S
NEWFOUNDLAND VOYAGE

In March 1886, Thoulet applied to his superior, the Director of Higher Education,[31] requesting a leave of absence "without loss of salary" and authorization to travel aboard a French naval vessel in the waters off Newfoundland.[32] He planned to extend his laboratory research and carry out a number of experiments analysing the topography and geology of the seafloor of the Grand Banks and the portion of the island of Newfoundland known as the French Shore.[33] According to Thoulet, French scientists had thus far been limiting their studies of the sea to marine zoology and largely ignoring *océanographie pure* – physics, chemistry, and geology – that is, "physical" oceanography.[34] He cited a number of problems relating to what he called the applied sciences of meteorology and hydrography, as well as his own work in "experimental" geology conducted in the laboratory. Arguing that it was vitally important to advance scientific knowledge of the world's oceans, he made a strong case that he ought to be enabled to undertake an expedition to Newfoundland waters, where according to the Treaty of Utrecht signed in 1713 by France and Great Britain, the French still had fishing rights, and where the French navy maintained an official presence six months of each year.[35]

After Thoulet made his application, a number of letters[36] were exchanged between authorities at the University of Nancy and the appropriate government officials in the Ministry of Education and the Ministry for the Navy and the Colonies concerning Thoulet's request for permission to travel as a civilian (specifically, as a functionary) on board a French navy vessel. Approval from the minister for what was obviously recognized as a worthy cause evidently posed no problem and was readily granted. There was, however, one matter that needed to be resolved: Who would pay for Thoulet's meals during the six-month journey? This was the topic of several letters and a potential deal breaker for Thoulet. The Minister for the Navy and the Colonies declared that his budget could not cover such expenses, and he wrote to his colleague the Minister of Education that he expected the Department of Education to pay the bill. The Minister of Education replied that his department was unable to absorb the cost, and he informed the *recteur* – superintendant – in Nancy that Thoulet would be expected to pay his own way.[37] When informed that he would have to bear all the costs for the entire six-month stay on board, Thoulet replied immediately that he had received no funding to cover expenses and that even if his meals were provided free of change, the cost of purchasing and transporting his scientific instruments was practically beyond his means. He asked the Superintendent of Education to intervene on his behalf with a request to the Minister for the Navy to waive any charge for his meals. Otherwise, he would be forced to abandon an expedition that was "hydrographical in nature and of particular interest to the Navy."[38]

Thoulet had the support of L. Grandin, the Dean of Science at the University of Nancy, who "in the name of science" intervened again on Thoulet's behalf and appealed to the Minister of Education, who in turn made a forceful plea to his naval colleague to reconsider his decision.[39] On 9 April, word came from the Minister of Education in Paris that the Minister for the Navy and the Colonies had generously agreed in this "exceptional circumstance" to waive the cost for Thoulet.[40] He was invited to present himself aboard the *Clorinde* in time for its May Day sailing from the Breton port of Lorient. Over the following six months, the *Clorinde* would sail many thousands of kilometres across the Atlantic, up and down the west coast of Newfoundland and around the Northern Peninsula, entering many of its bays, coves, and harbours. It would also visit Cape Breton Island and St-Pierre before returning to France. Thoulet would be able to carry out his scientific work and, moreover, to write down many and various personal observations about almost everything he saw and thought about. Those observations would become the subject matter for this book.

THE VOYAGE OF THE *CLORINDE* IN 1886

Thoulet would describe the *Clorinde* as "solid, seaworthy and comfortable ... a wooden-hulled frigate armed with twenty cannon, built as a sailing ship."[41] Records in the Service historique de la Marine in Vincennes and the Musée de la Marine in Paris indicate that the *croiseur à batteries* or armed cruiser was completed in 1845 in Cherbourg, her home port until she was decommissioned in 1911. Her displacement was 1,853 tons.[42] She had undergone major renovations in 1870 – in particular, she had been fitted with a 563-horsepower coal-fired steam engine, which, as Thoulet writes, would be "of great assistance, though it will never produce great speed."[43] (The ship's top speed was only 8.8 knots.[44]) Its captain was Félix Auguste Le Clerc,[45] who had already served on it for several seasons in Newfoundland waters, and records indicate that it carried twenty-two cannon.[46] She was to serve for at least eight seasons in Newfoundland waters, from 1880 to 1887. During 1885 and 1886,[47] she carried the flag of the *station navale* of the French fishery in Newfoundland, leading the *Crocodile* and the *Ibis*[48] the year before Thoulet's voyage and the *Drac*, 4 cannon, and the *Perle*, two cannon, in 1886.[49] According to Thoulet, during the 1886 Newfoundland campaign, there were two hundred men on board.

The *Clorinde*, now more than forty years old, was nearing the end of her active career on the French Shore. According to French naval correspondence,[50] after the 1886 Newfoundland voyage, orders were issued for repairs to be carried out. She had been damaged during an October storm in the mid-Atlantic while she was returning home (a storm related in vivid detail by Thoulet).[51] The work was ordered to be done immediately so that the ship would be ready for the departure of the naval patrol fleet to Newfoundland the following summer. The *Clorinde* did, in fact, return in 1887 – her last year of service in Newfoundland waters.[52]

Thoulet was no naval officer. Even so, his account of his voyage to Newfoundland is surprisingly short of the kind of precise chronological detail one might expect to find in a nineteenth-century travel narrative, especially one written by a keen observer on a scientific expedition aboard an official vessel of the French navy. At the beginning of *Un voyage à Terre-Neuve*, Thoulet states that he arrived in Lorient in late April 1886 and that the *Clorinde* set sail on 2 May.[53] Another of his works, "Observations faites à Terre-Neuve à bord de la frégate *La Clorinde* pendant la campagne de 1886,"[54] is more official in tone – not surprisingly, since it was Thoulet's official report on what he described somewhat grandiosely as the "mission" to which he had been entrusted by the Department of Education. Here, he indicated the dates of his Newfoundland expedition: 2 May to 22 October 1886. Yet in *Voyage to Newfoundland,* the days, weeks, and months go by, and he fills the pages with observations of all sorts, but only rarely is he more specific than "the next day," or "a few days later," or "one fine morning," or "autumn had arrived." Efforts to track as nearly as possible the itinerary of Thoulet's travels in Newfoundland by referring to the days he mentions having spent in each place have proven unsuccessful.

In one publication, however, Thoulet does supply precise information regarding where he was and when. In this article, he presents and analyses the detailed results of his measurements of seawater density and considers ocean currents around the island of Newfoundland.[55] Nearly ten pages of charts[56] list the water density and – for each entry – the date and the ship's position (indicated in latitude and longitude or in nautical miles north, east, south, or west of a landmark). He also records water and air temperature, barometric pressure, wind direction, and other observations such as the colour of the sea[57] ("indigo," "green," "grey," "greyish-green"), "ship at anchor," "halo around the moon," "dead calm," "jellyfish surrounding the ship," "dense fog," and "ship surrounded by icebergs." The first of these readings is for 16 May, when the *Clorinde* was approaching North America; the final one is for 14 October, when the ship was nearing France on its return. Yet even these charts indicate a number of gaps where the dates and corresponding positions have been omitted – presumably, when no measurements were taken nor entries made. My attempts to obtain as complete a record as possible of the ship's movements by locating the 1886 logbook of the *Clorinde* in the Service historique de la marine at Vincennes,[58] the Archives nationales in Paris, or the archives in Lorient (where the voyage began), in Brest (where the vessel ended her 1886 voyage), or in Cherbourg (her home port) have failed.

A plausible explanation for the absence of chronological detail can be found in an entertaining and well-illustrated account of the same voyage written by Louis Koenig, French naval lieutenant and officer on board, undoubtedly the person whom Thoulet refers to as "my friend K."[59] He describes Thoulet as a "professor of Natural Science, who was a charming and valuable companion" during the voyage. Koenig states why he himself made no attempt to provide a chronological account or log of the *Clorinde's* itinerary: "It would be fastidious to impose upon the reader ... the day-by-day peregrination of the *Clorinde* along the French Shore. The ship was required, by the nature of its work, to go from one bay to another, then often to return again to places already visited. The description of such an itinerary would bring confusion, from a geographic point of view, to the mind of the reader."[60] This sort of back-and-forth movement along the coast did allow for the possibility of dropping off livestock, specifically sheep, so that they could graze and grow fatter before being picked up some weeks later; and for a limited amount of planting of certain crops, such as radishes and cress, which would be harvested later in the summer as fresh produce to supplement the crew's diet.[61] The map I have provided shows only the places Thoulet refers to in his narrative; they do not mention the date or dates of his visits.

THE PUBLICATION OF *UN VOYAGE À TERRE-NEUVE*

Thoulet's account of his *Clorinde* expedition was first published in six instalments of about thirty pages each in the quarterly *Bulletin de la Société de géographie de l'Est*, from 1890 to 1892, and was very likely destined for a wide readership of nonspecialists.[62]

In 1891, five years after he completed his journey, Thoulet published *Un Voyage à Terre-Neuve,* with Berger-Levrault in Paris and Nancy. It consisted of one volume in octavo of 175 pages and carried an indication on the title page that it was an "excerpt from the *Bulletin de la Société de géographie de l'Est.* "[63]

The book was illustrated with four plates from Thoulet's own black-and-white negatives,[64] printed as phototypes by J. Royer in Nancy. An amateur photographer, Thoulet had made a series of photographs of places and people during his 1886 *Clorinde* voyage. He used these to illustrate the various presentations he gave soon after his return to France. Thoulet was not a member of the Académie des Sciences de Paris; even so, he sent a paper to be read or presented by that body's permanent secretary, Marcelin Berthelot, on 22 November 1886: "On the Formation of the Banks of Newfoundland."[65] Another of Thoulet's papers was presented to the same group on 13 December: "On the Erosion of Rocks by the Combined Action of the Sea and Coastal Ice."[66] On 18 February 1887, Thoulet gave a presentation at 8:30 p.m. at the Hôtel de la Société de Géographie, 184 boulevard Saint-Germain, Paris, with the following program: "A Summer along the French Shore of the Island of Newfoundland, Saint-Pierre et Miquelon, the West Coast, the North Coast and the East Coast (Eastern Shore of the Northern Peninsula) of Newfoundland, the Cod Fishery, Sydney, and Cape Breton Island." The projection was done by Mr Molteni, the geographic society's usual projectionist, using oxyhydrogen lamps (similar to limelights). On 8 May 1887, Thoulet made another presentation about his voyage to Newfoundland, this time in Nancy, to the Société de géographie de l'Est.

The photographs were collected in an album titled "42 photographies de Saint-Pierre, de Cap-Breton, de Terre-Neuve et du Labrador,"[67] which Thoulet donated in 1887 to the geographic society in Paris, where they are housed today.[68] Four of the photographs were chosen to illustrate the Berger–Levrault 1891 edition. Two of them – "Les cales près du cap à l'aigle" (The Docks near Cap à l'Aigle) (the frontispiece) and "Le quai de la Roncière" (The Roncière Wharf) (between pages 30 and 31) – show St-Pierre. Possibly, these were considered of greatest interest to French readers. A third photograph – "Roche érodée, baie d'Ingornachoix" (Eroded Rock, Port Saunders) (between pages 76 and 77) – is of Newfoundland. A fourth – "Main Street à South Sydney" (between pages 144 and 145) shows Cape Breton. In his account of the voyage, Thoulet mentions that he also made sketches,[69] in particular of the region of Cat Arm. These have not been located.

Travel by sea in the 1880s by someone in Thoulet's position undoubtedly provided the time and comfort to allow for the leisurely activity of writing. Although Thoulet claims that his "only preoccupation" during his voyage to Newfoundland was the pursuit of science,[70] his book is filled with observations and musings on a wide variety of topics, in keeping with the slow pace of the *Clorinde*. These sometimes arise unexpectedly; more often, they are inspired by the sight of some object of interest or curiosity that leads to a discussion of matters social, historical, political, sociological, literary, or pedagogical. A number of times, Thoulet writes about his work. As a scientist on an important field trip, his main objectives included studying and recording his observations on the mineralogy and geological composition of rock formations in northern Newfoundland. However, he was also writing a travel account, a description of his journey, which would later become this book, and he was evidently writing for a fairly broad readership, not only for specialists.

Thoulet's prose is grammatically flawless, and his vocabulary is rich and precise. His style suggests what his lectures must have sounded like. One of his colleagues at the University of Nancy, the geographer Bertrand Auerbach, wrote the following remarks: "The lectures in which Monsieur Thoulet explained to us the latest discoveries and thinking, were a delight; their clarity, high literary standard, spontaneous liveliness, all of which was enhanced by documentation and illustrations in the form of photographs and maps, made quite an original and unique collection."[71] It is also worth mentioning – although it may be as much of a reflection on the quality of the publisher – that the original *Voyage à Terre-Neuve* is entirely free of typographical and other errors.

Thoulet's book has twelve chapters; each corresponds (most of the time) to a lengthy or noteworthy stop made by the *Clorinde* (such as "The Islands of St Pierre & Miquelon"), to a major portion of the journey ("The Crossing" or "The Trip Home"), or to an important topic on which he has chosen to write at some length ("A Little Geography and History" or "The Cod Fishery"). The vessel's arrival in each new bay or harbour along its journey typically offered Thoulet a convenient and sometimes striking starting point for each chapter, and a logical beginning was to describe the mineralogical, geological, and topological features of the scenery.

On the very first page, Thoulet introduces himself and makes it clear that the purpose of his voyage is mainly scientific. Although the ship's actual sailing has been delayed a few days, he immediately starts the narration with a description of the town of Lorient and its surroundings, the walks he takes in the area, and his impressions. The anchor has not yet been raised, but the journey – by Thoulet's account – has begun.

The bays, coves, and harbours Thoulet visits and the nearby geographic features are his "official" priorities and the subjects of his scientific scrutiny. But as a traveller, he is all eyes and ears – his interests and commentaries are not at all circumscribed by the research he has set out to do. He subjects the landscapes, seascapes, and local inhabitants to his careful inspection as well as to his sketchbook or camera. The scenery has a strong impact on Thoulet the scientist; indeed, his many superlatives, which I have translated as *marvellous, extraordinarily majestic, charming, astonishing, terrifying, superb, admirable,* and *magnificent,* indicate that throughout his journey he responded to the sights with intense emotion. He is moved profoundly by the sight of shipwrecks on the isthmus of Langlade–Miquelon, known as the "graveyard of the Atlantic." He writes of the "impression of gloom" and of the "profound horror" produced by the "long reddish-black stains [on] the sides of the shipwrecks, like blood from a wound," and he likens the broken masts and spars to "huge disjointed limbs," and the entire area to "a vast ossuary."[72]

More than once in northern Newfoundland, Thoulet is moved by the natural surroundings; the following description of a sunset is so strikingly accurate that I reproduce it here in its entirety:

In this region of the country, winter must certainly be marvellously beautiful; but this evening, leaning over the rail, I admire the huge, red disk of the setting sun. The sea is a very bright shade of bluish green; small waves lap against the side

of the frigate. The light shines on the delicate outline of the
fir trees which stand out against the sky, where scattered purplish clouds slowly darken the upper portion while, near the
horizon, oblique pink and green bands shine with the brilliance of stained glass in a church. The masts of a schooner
at anchor, whose hull is lost in shadow, stand as straight and
taut as wires. The sun goes lower, the sea becomes steel blue,
the sky takes on the same shade as the sea, then becomes
purple; the metallic colour of the water fades and the blue
deepens as darkness begins to fall, first leaving a thinning ray
of gold, now only a burning trace soon to be extinguished
while the cool evening air spreads itself around; the winds
dies; night has fallen, and the earth, the sea, and the sky now
sleep in silence.[73]

Reading the following passage, from near the beginning of
chapter 11, one is reminded of two nineteenth-century French
poets, Baudelaire and Rimbaud. "Names carry with them a certain rhythm, a musical quality which makes us like the objects or
the creature which bear them even before we know them, hate
them, or feel indifferent towards them … The combination of syllables in the name Iceland evokes the idea of a dark land, full of
sublime horror; by simply pronouncing it, you hear the sound of
a block of ice cracking; the name Spitzbergen makes you think of
a desolate island, with rugged mountains and sharp peaks reaching up through the fog; when you say Novaya Zemlya,[74] you
dream of grey skies and black rocks protruding through the
snow."[75] Thoulet associates colours and sounds in a manner that
is reminiscent of Baudelaire, who in "Correspondences"[76] wrote:
"Perfumes, colours and sounds reply to one another." Thoulet,
like Rimbaud in "Voyelles,"[77] writes that the syllables that compose
a name are a source of musical and visually colourful impressions.

The geology of the regions he visited was of course immediately clear to Thoulet's expert eye. His account sometimes takes
on the professorial tone he would have used to define and explain to his students the features he observed, such as the *tolts*, or
peaks of the mountains of the Northern Peninsula, and the indentations along the Newfoundland coast known as *fjords*.

Thoulet's account is not a scientific treatise; even so, it will occasionally offer today's readers a glimpse of an earlier stage of
scientific knowledge, before geologists fully understood the fun-

damental processes of formation and transformation of some of
the major features of the earth's crust. Some of Thoulet's technical explanations are convincing to a nonspecialist; indeed,
they often turn out to be partially if not entirely accurate. Others
belong to outdated nineteenth-century theories. For example,
he refers to the formation of fjords as "one of the least disputed
questions of geology"; he declares that they originated in the
quaternary (Ice Age) period, and notes that a fjord is "a sign that
the terrain has sunken."[78] In fact, it is now known that they were
formed when the Ice Age glaciers melted and sea levels rose.

As Thoulet indicated in his initial letter requesting authorization to travel to Newfoundland, two particular aspects of geology that he specifically wanted to study were the effects of
frost and ice on the erosion of rock formations along the coast
and the geological composition of the offshore banks.[79] Regarding the latter, he questioned the scientific thinking of the
time, which, based on analyses of sediment samples taken from
the seafloor at different depths, attributed the formation of the
Grand Banks[80] to an accumulation of sediments that had been
transported there by icebergs and deposited where the ice melted – that is, in the zone where the cold Arctic Current meets the
warm Gulf Stream. Thoulet contended that icebergs could not
possibly have transported enough solid matter to contribute sig-
nificantly to this process.[81] His own hypothesis was that fragments broken off along the coast were picked up, transported
by shore ice, and subsequently deposited in the sea when the ice
melted, in the area we now know as the Continental Shelf.

In another example, Thoulet's description of the "immense
plateau of yellowish-red rock without any vegetation" refers to
the mantle rock in the region known as the Tablelands, which
are now part of Gros Morne National Park, which UNESCO has
declared a World Heritage Site. Thoulet speculated about its resemblance to what the landscape in Iceland or Greenland would
probably look like underneath a glacier.[82] Obviously, we no longer read Thoulet for his science; however, his contributions were
deemed useful by his contemporaries. One of them, the geographer R. Perret, visited Newfoundland and in 1913 published a
lengthy treatise on its geography.[83] Perret praised Thoulet's
work, referring to his geology of the Newfoundland Banks as
"doctrinal" and as "present in everyone's memory."[84]

THE PLIGHT OF THE FISHERMEN AND
THEIR BRAVERY

Readers who are familiar with the moratorium on the northern cod fishery in the 1990s and the dramatic impact this has had on rural Newfoundland will recognize some of the conditions Thoulet describes when he examines the fishery in the mid to late 1800s. In certain parts of Newfoundland, people were already being forced to leave their fishing communities and move elsewhere. He writes:

> If there are no cod to be found, the fishing settlements are abandoned, the stages fall to ruin and collapse and rot in the grass; there are no more boats moored, no dories in the harbours, no life anywhere. For the remaining sedentary fishermen who tried to outlast their bad luck [it means] solitude and misery.[85]

Thoulet takes what he calls a "fair look" at the difficulties encountered by the French fishermen in their relations with the English. Although it cannot have been easy for Thoulet to remain impartial – as shown by his remark about "the damage caused by narrow mesh nets used by the English fishermen"[86] – he seems to have managed reasonably well. Thoulet summarizes how France and England, although no longer at war after the mid-1900s, had signed a number of treaties and agreements but were unable to resolve their points of contention.[87] In chapter 4, "A Little Geography and History," he discusses relations between the two countries and the numerous and lengthy negotiations and subsequent failures.

Thoulet devotes all of chapter 9 to the cod fishery, describing its importance not only to Newfoundland's economy, but also to that of France. He explains the differences in the various techniques employed on the offshore vessels called bankers and schooners and by the dory fishermen of Miquelon and Île aux Marins. He discusses the relative difficulties and advantages of each fishing technique and the industries that have developed around fishing, such as the production of cod liver oil and recent innovations in methods for drying cod.

After almost a month on St-Pierre, Thoulet spent several months travelling along the French Shore. He gives a long and detailed description of lobster fishing, processing, and canning at Ingornachoix Bay and Port Saunders that is striking and sometimes funny. Excerpt:

> When I visited, there were three or four thousand lobsters laid out in this way. Such a colourful sight would undoubtedly inspire an impressionist painter, except that such a rubicund congregation, such a conclave of cardinals of the sea, of course all deceased, would necessitate a substantial expenditure on vermillion or Saturn red.[88]

Throughout the book, Thoulet recounts many of the conversations he had with the local people about the cod fishery. He goes on at length about the hardships the fisherman must endure, the low wages, and the dangers they face. Perhaps exaggerating, he writes: "If it were not for the fishery, there would likely have been little or no settlement in Newfoundland, and certainly little reason for France to be interested in the place."

For a number of reasons, it is logical that an oceanographer interested in the ocean – even one like Thoulet, whose specialty was physical oceanography – would be preoccupied with the fate of fishermen. In the early 1880s, a crisis developed in the sardine fishery on France's Atlantic coast. During the 1886 season that Thoulet spent in Newfoundland, Prince Albert of Monaco carried out his second scientific campaign aboard the *Hirondelle*. During his stays in the ports of Spanish Galicia, he compiled a detailed study of the conditions that prevailed in the sardine fishery, and the following year he published a lengthy treatise on the subject. It is worth noting as well that in 1886, Pierre Loti published *Pêcheur d'Islande*,[89] a novel – still popular – that described the unbelievably punitive conditions that Breton fishermen had to confront while fishing for cod off Iceland. At the time Thoulet was writing, the public knew a great deal about the miseries faced by fisherfolk.

At the time Thoulet was sailing with the *Clorinde,* ocean liners were already racing one another across the Atlantic. This presented a grave danger to the many tiny boats fishing on the Grand Banks, many of which had only one or two men aboard. This was long before radar; the dories, hidden by fog and rough seas, were often run over or swamped by much larger steamships, whose crews were oblivious to them. The danger of collision between boats of all sizes was a theme would recur more and more often after the middle of the nineteenth century.[90]

THE MYTH OF THE NEWFOUNDLAND DOG

Thoulet, always the pedagogue, set out to correct what he saw as a popular misconception regarding the Newfoundland dog, widely reputed – especially in the places he visited – as capable of the most heroic rescues:

> If ever I am in danger of drowning, God preserve me, I hope with all my heart that there is no Newfoundland dog nearby … They can save anything that floats, pieces of wood, algae, froth, and even a man, provided he does not move, because if he does, the dog will put his heavy paw on the head of the poor unfortunate, flailing about, and will keep him underwater until he drowns, then bring him ashore.[91]

Thoulet was a seasoned traveller whose curiosity was occasionally rewarded with the discovery of legends such as those behind names. He shares the origin of Tête-de-mort (Death's Head) Harbour, originally known as Maiden Bay, and recounts this moving story of loneliness, sickness, and death in chapter 11.

Place names provide many interesting and occasionally humorous facts, such as the following:

> In the United States, in the state of Iowa near Galena, there is a river called the Fever River, which was simply the rivière de Febves or Fèves [or the River of Beans] named by French explorers. When I was carrying out a topographical survey in northern Minnesota, I remember naming a lake after a Monsieur Maunoir, the likeable Secretary General of the Paris Geographical Society. The lake was one I had taken pains to choose among the most picturesque, in honour of its patron, but it had the misfortune of ending up being called Manure Lake![92]

THOULET'S CULTURAL AND LITERARY BACKGROUND

Julien Thoulet was a scientist, but his style of writing called into use his knowledge of both modern and classical literature. References to literary and artistic works are frequent in Thoulet's writings, and they tell a good deal about nineteenth-century schooling in France and his own education[93] – a vastly different cultural and literary tradition than what prevails in today's Newfoundland. I have therefore provided notes that attempt to clarify as many as possible of Thoulet's literary references and allusions, as well as contemporary names that are either vague or totally meaningless for most today's readers.

Although writing in French, Thoulet often resorts to English words, so he apparently spoke the language. His stay in the United States, "of several years duration,"[94] must have been sufficient time for him to learn it. Regarding his English skills, Baron Jules de Guerne, one of his main scientific collaborators at the time, offered the following precise description in a letter to Prince Albert: "Monsieur Thoulet will be free on July 11. You would do him a great service ... to bring him to Edinburgh. He speaks perfect English and would be very useful to you from that point of view."[95] His bibliography includes two articles with titles in English published in the United States.[96] Also, there is a reference in French to a work translated from the English, presumably by Thoulet himself.[97]

Thoulet more than succeeded in carrying out the work he had planned for his Newfoundland expedition. Besides taking scientific measurements and conducting experiments, he observed icebergs – he even tasted one![98] He studied the composition, density, and temperature of seawater, the erosion of coastal rocks, and the speed and temperature of surface and subsurface ocean currents; all of this contributed to the burgeoning science of oceanography, which Perret referred to as the geography of the sea. For Thoulet, a pioneer in this new field, the timing of the *Clorinde* voyage was perfect. It enabled him to begin in earnest the pursuit of what was to become his lifelong scientific quest to understand and explain the physical phenomena of the ocean. Long before his death in 1936, Thoulet was recognized as the father of French oceanography.[99] Prince Albert of Monaco made the following remarks in his opening speech to the Conférences sur l'océanographie in Paris in 1903:

> Today you will hear a scientist who has devoted himself to French oceanography and who, despite the obstacles in his way, has made himself quite a name among oceanographers from other countries where substantial means are employed to further this science. Professor Thoulet will interest you because he has often gone outside his laboratory and worked at sea, where the direct observation of facts provides the mind with a vision of more depth and precision.[100]

Thoulet's *Voyage* has never been declared a literary masterpiece; even so, it is a fascinating and well-written commentary about a place few people know. Nearly all of that region's history has been written from a British perspective. Thoulet's book, translated into English for the first time here, is the account of an educated and perceptive nineteenth-century French scientist who spent the summer of 1886 on the French Shore.

SCOTT JAMIESON

A Voyage to Newfoundland

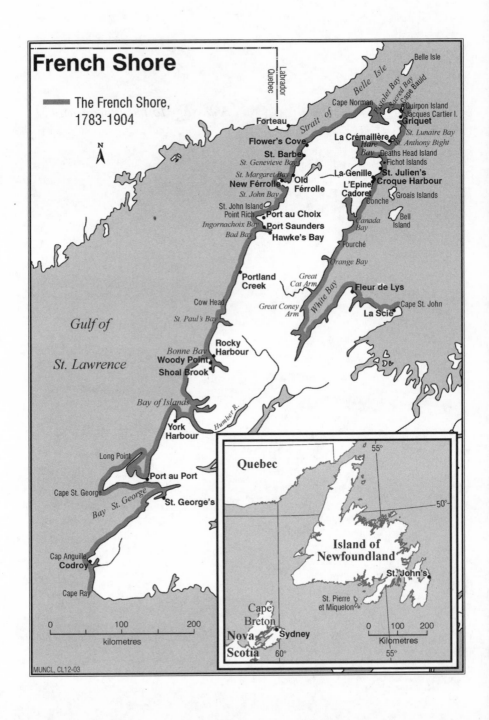

1
Lorient

I arrived in Lorient in late April 1886 to embark on a journey aboard the *Clorinde,* bound for Newfoundland, in order to carry out certain oceanographic observations and learned that, instead of departing two days later, the frigate would not sail for another week. However, I was promptly consoled – for the next six months, I would be living aboard ship and was therefore not at all displeased to have the time to bid a leisurely goodbye to dry land. The best part of any kind of joy is the beginning or the end. A gourmet would probably argue that what he prefers about a good dinner is the moment when he unfolds his napkin and takes his first mouthful of a full-bodied wine. One thing is certain: true happiness is in the anticipation of the days to come, and nothing is more enjoyable than awaiting departure when all the preparations are complete and one is neither hurried nor impatient. I went aboard the frigate; on deck, the men were dressed in grey, with blue woollen caps, white chinstraps, and red tassels. They rested on the spare sails or in the launch, sitting around on coils of rope or around the cannon, enjoying their Sunday rest by dozing or chatting in the sun. It took me very little time to get settled. As soon as my baggage was stored away in the cabin that would be my sleeping quarters, study, and laboratory, I went back on shore.

Lorient is a small, pretty town and can be seen and admired without having to spend long hours. You start from the train station, then cross the fortifications, follow rue Victor-Massé, named after an authentic Lorient hero,[1] and continue on around Place

Alsace-Lorraine, proceed along rue des Fontaines to place du Marché, next to the church, in front of the statue of Bisson, then along cours de la Bôve to arrive in front of the theatre. If this walk of some fifteen minutes is not enough to leave an impression, then all you have to do is to start over again in the opposite direction in order to be perfectly informed about all the town's beautiful sights. After the twentieth or the fiftieth time, the same walk becomes somewhat monotonous, so that you feel like going to admire the country rather than city sights. So I decided to leave on foot and set out alone for Hennebont. The road is magnificent and I am not daunted by the ten kilometres walk each way; moreover, I intend to shorten the distance by conversing with myself.

Each of us possesses within himself a delightful interlocutor. When one "me" proves too stupid or too clumsy, which does happen, alas, the other "me" is full of indulgence and excuses, even forgives the former's sullenness. If he is bothersome, the other smiles politely and gives him enough time to prepare all his replies, a minute, an hour, a day, a week if necessary. In the meantime, the topic of conversation changes, since there is no shortage of inspiration. A blade of grass, a stone, an ant walking busily, a passing cloud, and things always get interesting. When my number one "me" is clever and witty – which he occasionally is, provided he is given the time – my number two "me" understands him well, admires him sincerely, is grateful to him for his replies, and enjoys his success without the slightest jealousy. What enjoyable moments "we" spent together, walking along the hedge-lined roads or the sunken paths where I was sure to meet no other traveller.

In April, the countryside in Brittany often resembles a man's serious and sad face when it is suddenly lit up by a smile. The ditches are dotted with violets and daisies, and wherever water has collected they are full of gladiola. Oak trees are cloaked in bronze-coloured moss, decked with wreaths of ivy hanging from their branches, covering everything from last year's fallen leaves to this year's new shoots. Blooms are everywhere; the buds themselves are of such a fresh, delicate shade that they too look like flowers. In places, stands of dark-green pine dotted with a harsher green serve as a foreground to the yellow trees beyond. The road stretches onward beneath the bright rays, without the brutality of the July sun. It is cool and warm at the same time.

4

Walking, one feels healthy and cheerful, strong and full of enthusiasm. The festive world around us has an effect on us just as we do on it. Strong mutual ties connect man and nature: we speak to her and she to us; she gives to us and we to her. When we contemplate her, we are convinced that life signifies unity; we feel the unity, the sense of community; we breathe the same air as all beings, all plants, all stones, all animals, and all men. Such impressions are more readily felt than expressed, but how many things are there which we cannot describe but which nonetheless exist? A writer of genius is one who is able to express in words that which nearly everyone feels deep in his heart. The dying oak leaves are replaced by the new ones in the same way that sad memories make way for hope; the transformation from one to the other is not always as bitter as the melancholy poet would claim. We observe and take pleasure in all that surrounds us and seize the day.

After coming down a hill, you arrive at Hennebont, where a few small boats are tied up on the river. Opposite a bridge stands a castle with granite walls and machicolations, a terrace covered with white and violet lilacs, the cracks in the stone filled with clusters of purple *joubarbs*[2] and yellow flowers; beyond, there is an open-beamed belvedere with a slate roof and a wine store-room. To enter the town, you follow a street leading up the hill, past a wash house supplied with water from a rushing stream and shaded by tall trees. One breathes the pleasant scents of spring. Further along there is an old door framed by two towers like those of the Jersual Gate in Dinan. At the top of the hill is the church with its pointed steeple, pinnacles connected by flying buttresses, built of granite that the passage of time has covered with reddish-yellow lichen. Who thought of calling these poor fungi whose warm colours enhance the sombre shades of the stone "wall leprosy"? The worst is that the term is absolutely fitting, which proves that it is often wrong to be utterly precise. What is necessary above all is agreement between the thing that is observed and the eye that observes. The church is nearly empty, with only two Breton women in white headdresses counting their rosary beads. The sound of footsteps can be heard on the stone tiles, echoing off the vault, muffled by the oaken pulpit, the woodwork, and dark walls. The sights, the sounds heard, and the thoughts inspired emanate harmoniously, forming a vague overall impression like that felt on an evening spent beside a calm ocean or during the heat of the day beneath the foliage of a shaded forest. No matter what its nature, a sensation is like a vibration or a sound. Writing is of necessity brutal, painting captures less, sculpture still less, and music less again, which is probably why music is the most perfect of all the arts. In Hennebont, there is no dissonance. The houses, the streets, the sky, the inhabitants dressed in their dark clothes go about their business, all in minor key. Breton peasant women have a distinctive appearance. Thanks to the white headdress, black dresses with bodice and velvet-trimmed armholes, short, stiffly pleated skirts over black stockings, black shoes, the humblest woman's entire person possesses that indefinable characteristic we call distinction, which is so rare among the peasants of France that I doubt whether it can be found anywhere outside of Brittany. We may speak of hidden and dormant volcanoes; it seems that the heart is more placid than the face. Who knows? And besides, of what import is it to the passer-by?

Another pretty walk is Port Louis. You board a small steamer at the quay in Lorient and are then taken across the harbour to the village of Pen-Mané. You follow the seashore along rounded beaches covered with hard sand, eroded from the white mica granite. In places, a thick cluster of grass grows where the tide has deposited silt. On the other side of the beach, a few cows graze in the swampy fields bordered by gorse; trees are scarce and grow close together to better resist onshore winds. In the distance, beyond the harbour entrance, lies the island of Groix. To the left is Gâvre and its arsenal, whose cannon can be heard. To the right is Larmor, facing Port Louis surrounded by its walls. The town resembles St-Malo – *si parva licet componere magnis*,[3] though the last word is perhaps rash in reference to St-Malo – without its memories of the past and especially without its lively streets. There is a stale, moral odour, the Louis XIV smell of certain ancient pieces of furniture long forgotten in some attic. St-Malo is suffocating within its old walls; as for Port Louis, which misfortune has befallen, it has become asphyxiated; Port Louis is dead. Some contend that there are still people there who manage to be happy and unhappy, to sleep and stay awake. That is extraordinary; let us bow our heads and bless Providence.

On Easter Sunday, there is a fair in Lorient. Place Alsace-Lorraine and the quays are filled with stalls. Country women walk in groups, their shoes bearing broad silver buckles, on their heads a cylindrical tiara covered with black material in tight folds radiating outwards, falling to the shoulders. The men have kept only the round hat trimmed with velvet ribbon. I dare not undertake to give a funeral oration on the demise of past customs and costumes; such lamentations have already been sung over and over. I occasionally take one up again, but fortunately for my own strictly personal use. I console myself with the thought that bygone days, falling leaves, and old memories go the way of all outdated notions and that, after all, our children still possess the Kingdom of the Congo and, thanks to science, the other planets will probably be at the disposal of our great-grandchildren.

Arab music can be heard coming from behind a canvas in one of the stalls, so I enter. An Algerian Jew, blind in one eye, with a pockmarked face, shoulder-length hair under a greasy skullcap, dirty clothes, and worn-out slippers, crouches in one corner of the stage strumming chords on a *gumbri* and occasionally scratching himself. On the other side of the stage is a second performer, small, shifty, and as pockmarked as the first one, with a broken tooth. In back, there are four women: one Negress and three Jews – the "Moslem Princesses" presented on the poster – with loose pantaloons and painted faces, sitting on cushions that are not of African origin. Each one rises in turn, comes in front of the spectators, dances and waves a cotton scarf. Occasionally, the musicians shout and the choir of dancers joins in, while from outside comes the voice of the Maltese impresario, whose moustache and beard are trimmed short, squarely, and well above the lips. Inside there is a dreadful smell of sweat. This style of music, with its shrill singing, strident accompaniment of the two strings of the *gumbri*, monotonous droning of the *darbouka* drum, I had once heard in Algiers, when it was still Algiers, when there were still Moorish cafés in the narrow, arched streets. The cool darkness was dimly lit by a small hanging night lamp, with the tiny red cinders covered with ash where the *kaouadji* heated the coffee he sold for one *sou* a cup. One sat cross-legged on a wide bench that was attached to the wall and covered with a mat. There were few customers, hardly a half-dozen. Sometimes, two of them played chess with roughly carved pieces on a wooden board with alternating indented and flat squares. You smoked without speaking, everyone listening to the silence. Suddenly a voice would begin to sing very softly, an endless murmur, with the muffled rhythm of fingers tapping on the skin of the *darbouka*. One would remain still, intoxicated with repose, eyes wide open, staring far, far into the distance at things in the sweet land of dreams.

The man next to me pushes with his elbow, the small Jew rises, does a short dance with his tongue out, then kneels down and, kissing the stage, bowing and rising, recites the Moslem prayer. He performs a parody of the gestures of someone praying to Allah, his unclean mouth repeating and sniggering at words meant to be sacred, no matter what the beliefs of those pronouncing them, words used every day by the weak to invoke strength and justice, by the unfortunate to seek consolation, by the desperate to plead for hope. The spectators burst out laughing and call out their intense satisfaction; the performer continues his prayer, raising his arms to the sky. When he bows, his skullcap falls and the crowd roars with laughter, taps its feet with delight while I quickly leave and go to sit quietly at the end of the wharf. I look at the hills surrounding the harbour, Pen-mané, Port Louis, Larmour, and the huge *Clorinde*, motionless at her mooring. Sparrows, with the boldness of spring, come to fight at my very feet. The sea sparkling in the sun and the monotonous sound of the rising tide lapping against the pilings of the wharf whisper softly, telling me that each one takes his pleasure wherever he can find it and that, for me, there is no need to look for it at the Lorient fair.

2
The Crossing

On Sunday, May 2nd, 1886, at eleven thirty o'clock in the morning, we sailed. Since we were simply made fast to a mooring, all that was needed was to let go the lines. The engine started and the frigate began to cut through the waves of the harbour. I could look through a scuttle in the gun deck near where I stood to avoid getting in the way of manoeuvres and see the trees, houses, ships at anchor pass slowly at first, then more quickly as we gained speed. When a man walks, he has the impression of remaining still while the objects around him seem to flee. So it is in life as we advance in age; our hair turns white and everything about us changes, our character, our tastes, our body, and our mind, and yet we think we are still the same. When we accuse others of changing and becoming strangers to us, we do so almost in good faith. Alas, we are the ones who change as we are carried off faster and faster in the whirlwind of time.

The *Clorinde* left Saint-Louis on our port side and soon fired a three-cannon salute to the Larmor church, which answered with a chime of its bells, an echo of the voices of those we left behind and a final wish for a good trip. Such salutes are a tradition in Lorient, and no ship leaving port would consider ignoring it because Our Lady of Lorient is jealous of the homage sailors pay her; all those who refuse are severely punished. During the Crimean War, two ships did: the *Pandour* and the *Sémillant* forgot or neglected to salute her, and subsequently, the former was lost without a trace and the *Sémillant* was crushed on the rocks in the

La Clorinde, Lorient 1880. © Musée de la Marine

Strait of Bonifacio,[1] and all on board were lost. Next we passed Groix, where a white semaphore stood out on the tip of the island. Its colour made it possible to distinguish it on the horizon for quite a long time; it gradually became a dot, smaller than a seagull flying above the waves, then finally disappeared entirely. We will not see France again for six months. The sea was calm, the sky blue; the first step taken on a trip is also the first step on the return home.

The *Clorinde* is a wooden-hulled frigate, armed with twenty cannon, built as a sailing ship; in spite of her massive size, her slender masts give an overall impression of elegance. When all her sails are flying, she is beautiful and does justice to the comparison with a galloping horse. An engine has been added and will be of great assistance, though in spite of all its best efforts, it will never produce tremendous speed. There is no comparison with the steamers crossing between Ireland and New York, whose beams cut the waves like a knife, indifferent to calm seas, wind, or storm. However, our ship is solid, seaworthy and comfortable; such advantages are not to be made light of. Moreover, the *Clorinde* knows Newfoundland. Every summer for the last few years, she has flown the colours of the commander of French naval surveillance over the cod fishery in Newfoundland waters.

The rolling of the ship, the rhythmic movement from left to right, or to use sailors' vocabulary, from port to starboard and back to port again, and the pitching movement of the full length of the ship from front to back are among the dangers that must be faced from the very outset of the journey by carefully securing all objects, instruments, books, whatever effects are kept in the narrow cabins on board. Very often a roll produced by one of those huge, silent swells from the open sea suddenly and stealthily tips the ship, and immediately the clattering of breaking dishes alerts us to the damage done. You hurry down to your cabin only to be taken aback by the sight that meets your eyes. You pay dearly for your lack of foresight; a pair of boots has broken the chamber pot, the mirror has fallen from the wall, the fishing rod has fallen onto the bunk, a pile of paper has slid behind the furniture, the inkpot has tipped over, the books have tumbled down on top of one another, and, since the *armoire* was left open, all the clothes and linens it contained are now on the floor. It sometimes happens that an exceptionally strong wave comes in through the scuttle and floods the entire cabin, producing what is called a *baleine*.[2] May God protect us from whales! Everything is soaked; you are lucky indeed if the bunk is not sodden in the process. After all your mopping, sponging, wiping up all the water, the dampness remains and dries very slowly at sea. I myself need not fear an accident of that sort. As soon as we sailed, the porthole which lets light into my small cabin was firmly shut, and its round opening, covered by a thick glass lens protected by a brass ring, was greased with tallow and attached on the inside by a hook held in place by an iron clamp. Just as the lion on a cobbler's sign can tear apart a boot but not break the stitches, the waves may break the glass of my porthole – though I certainly hope that it never happens – but they cannot seep in through the smallest opening. If by misfortune a drop does penetrate, it would fall into a large funnel beneath the porthole and drain into the sea.

My cabin is located on the port side of the orlop deck, that is to say, below the gun deck, which is itself below the deck, which is in turn below the poop deck so that whenever I wish to look out at the sea I must climb three decks. The captain has his quarters below the poop, on the deck level; next to him are the first officer, the second officer, the five lieutenants, the ship's doctor, and the purser, each with his own quarters on the gun deck, at the end of which is located the officers' mess. The midshipmen's quarters and the stewards' cabins are on the orlop deck. What is done at sea is the opposite of what happens on land, and according to a certain logical principle, the higher the rank, the higher up the living quarters. On board ship, you enter from the top, just as wine enters through the top of the bottle.

My cabin contains a bunk which takes up the entire width. Against the bunk, there is a set of shelves that is my laboratory. Next to it, there is a chest which serves as a wash stand whose top drawer can be transformed into a table or desk, and finally a clothes closet. The third wall of the cabin is taken up by the door, which can be held open by a hook. By opening a red curtain, you can breathe more air. Against the fourth wall is the *armoire*, along with a box, two trunks, and three jugs of seawater for my experiments and, finally, the one and only chair. I am able to accommodate one visitor, but if two arrive, one must remain outside. Near the ceiling, there are shelves for books and various utensils. On the bulkheads there are several pegs, a lamp, a photographic lantern, a mirror, and a barometer. The room is lit by the porthole situated at the end of a cylindrical opening the thickness of the ship's hull. At sea, there is no overabundance of fresh air or light.

My days on board pass with perfect regularity. I awake at the reveille to the sound of the morning bugle and drum, get up and go to the mess for breakfast. I begin working, carry out my observations, eat again, resume my tasks, eat dinner, spend the evening in the mess, retire at an early hour, and fall asleep to the sound of the water running along the hull of the ship, since my cabin is just above the water line. Every tomorrow resembles every yesterday.

The mess room! This word sums up life on board for the officers. I refer, of course, to their private lives during off-duty hours. The captain has his own quarters and takes his meals with the first officer. He walks around the outside balcony astern and rarely comes on deck. The midshipmen live, eat, sleep, sing, talk, read, dress, and undress in their quarters. Each officer has his bunk,

but they all meet and eat together in the mess. The *Clorinde's* mess
is huge; it occupies the stern of the ship and is semicircular in
shape. It is painted white with a long divan of red repp[3] along all
but one of its bulkheads and is lit by three scuttles. In the middle
of the room there is a large oval table. From breakfast until the
end of the evening, this room serves as salon and dining room
and is continuously in use. Occasionally, when it rains and the
wind is unfavourable, everyone stretches out on the cushions and
takes a nap. On other days, the chance choice of a word provokes
one of the repertoire of standard and numbered discussions that
take place in the messes of the French Navy. For example, discussion number fourteen: Is Food Better in the South of France or
the North? Alas, whether the poor officers originate from Toulon,
Rochefort, Lorient, Brest, Cherbourg, Senegal, or China, they eat
poorly just about everywhere. It is unusual for cooks on board,
who are civilians hired for the duration of the voyage, to possess
the many talents necessary: cook, baker, and pastry chef. They
often take advantage of this circumstance once the ship is at sea
and their services are indispensable. Journeys are long and fresh
provisions in short supply. Where are the good home-cooked
meals? Best of all, where is that delicious taste of roast mutton
which Victor Jacquemont[4] longed for while in the Himalayan
Mountains? We talk about absent comrades whom we sailed with
for a year or two, shared our lives with, slept next door to, sat at
the same table with, ran the same risks with, and experienced the
same joys and fears, the same impatience. Now they are scattered
all around the world. Occasionally, someone relates events that
he witnessed or something he accomplished himself, and the real
story bears little resemblance to the one told in history books or
especially in newspapers. We discuss music and, for want of another instrument, mimic the gestures and hum the airs from old
operas and operettas we remember. Meanwhile, others play cards,
dominoes, backgammon, patience – which some call the consolation of broken hearts – and cup and ball, the only game that can
be played at sea without concern for the ship's rolling and pitching and which can inspire highly skilled variations, the fruit of
diligent practice.

All things considered, the life of a sailor is an unnatural one;
everything is intertwined: sea and land, departure and return, the
heat of the tropics and the cold of the poles. Throughout this irregular existence, you come to esteem and like one another, and

then you part company or you hate one another and end up back together. One thing only persists, the most difficult of all, and that is absence. While the sailor looks at the passing waves, day after day, back home, others live and die, events carry on in his family and in his country without his taking part; the only joys and sorrows he can feel are dulled. To be a sailor, one must be young. The journeys hold their charms perhaps until the age of thirty, then the continuous novelty soon becomes terribly monotonous. Later, the burden becomes heavy. How many continue their career because they no longer dare or are no longer able to exchange it for another? And try as one might to heap the honour upon this life which it so rightly reserves, to make it seem sweeter by whatever means available to modern civilization with all its progress, it will always remain false, because man was not created to live alone. No matter how great his courage, his abnegation, his enthusiasm, his ambition, he is incapable of depriving his heart, his mind, and his body of that which they were made for and which they demand so imperiously, without suffering the consequences, at a time when youth casts its exuberance to the wind and age begins to impose its rights. The sea is a life for the young, and one does not remain young very long.

I write these lines without bitterness, of course. I am not a sailor, yet I have been welcomed most warmly by sailors during the time I have spent in their company; they have bestowed such attention upon me, received me so cordially, with such thoughtfulness, and so often paid me the honour of treating me almost as one of their own that I have a deep memory of gratitude and sincere affection toward them. I would be in utter despair if any of them were to see in the expression of my thoughts the slightest intention to wound. Can it be considered as hurtful toward them to view their task as men as a superhuman one? With the exception of a few particularly sturdy characters, most of them pay very dearly during their mature years for their dreams of sparkling skies, battles, beautiful dark eyes, storms, coconut palms, and the hot sun so well described in the novels that fired their young imaginations.

The frigate continues on its course. The weather is magnificent, the sea made rough by the majestic swell which can be seen arriving from the horizon; it is like the peaceful and rhythmic breathing of the ocean. We are heading for the Azores in search of the trade winds. The weather is getting hotter and the water has taken on the most striking deep blue, and each evening the sun setting behind the golden clouds paints a wide red band. The moon rises and speckles the waves with glitter; the sea and sky, studded with stars, blend into one, and the ship leaves a phosphorescent stream in its wake. Jellyfish float around us, their gelatinous bodies raised up into the wind like sails, trimmed purple and trailed by a bouquet of long filaments, infinitely varied in colour and form or like clusters of tropical grapes. Birds glide along behind us, gulls and petrels, flying continuously from one side of our wake to the other, tireless travellers to whom we say adieu in the evening when we go down to our bunks and whom we see again in the morning when we climb up once more to the poop deck. Sometimes, a poor little bird, lost in the middle of all that immensity, notices a speck in the distance, and that speck is us. It expends the last energy its exhausted wings possess and comes to rest on our main yard, where it hesitates a long time before alighting. Its fatigue is so extreme that it falls and dies, or else, it stops for a moment, rests, and then flies away again toward its unknown destination.

We sail on and head north. The wind freshens, the seas swell, and wave follows wave. The ship leans and the cattle tied on deck – becoming fewer and fewer as they serve as food for the crew – stretch the ropes that attach them to the rail and brace their hooves against the deck as it seems to slide from under them. I am sitting in my cabin with my pen and paper before me while my porthole plunges below the water line and the light coming in through the glass casts a dull bluish-green on all the objects around me. A large cover is placed over the table; in it are holes in which pegs are placed to prevent plates, glasses, carafes from sliding. It is almost a pity the floor itself does not have similar pegs, because it often happens that during a meal, a person's chair leaves the table and slides to the other end of the mess. Precautions must be taken such as eating with one hand and holding on with the other, only half filling one's plate; otherwise the soup ends up on your lap. To avoid a somewhat colder but equally unpleasant flooding, you must pour only the amount of liquid that you can drink in one mouthful. At night, you must lie against the board at the edge of your bed; you fall asleep if and however you can. After a night of heavy rolling, you are lucky to get off with only a stiff back. How happy you feel to leave the narrow cabin and go up on deck to breathe the fresh morning air; whether it is raining or windy you feel invigorated by the coolness and happily await scrubbing time.

Scrubbing the deck is the joy and glory of the crew and misery for the passengers. The men are bare legged and wash abundantly, the pump working energetically since there is no risk of the reservoir running dry. The decks are flooded and brushed with brooms, scrubbed with soap and sand, mopped dry, scraped with iron scrapers, and finally mopped dry again, and the first part of the operation is finished. To empty a bucket of water and spread the contents evenly over a large area by a circular movement of the body, throwing the water forward, is an art which requires talent and is a privilege, indeed almost a calling, reserved for officers only. That is not all: the brass, the hinges, the cannon, the guns in their racks, which are kept under canvas except for Sunday inspection and admiration, all have to be polished and shined. The opium smoker inhales his smoke and enters into a state of bliss; the Hindu fakir contemplates his navel and sees the sky open up; the sailor polishes and his eyes gaze into infinity; he shines and daydreams and, in his thoughts, escapes from the authority of the quartermaster, the second officer, the first officer, the master-at-arms, the terrible keepers of the records. The sailor shines and shines and shines and is happy.

A roll of the drum indicates that the washing of the deck is complete. Next the series of daily exercises begins, to be interrupted by the second meal, which takes place at ten o'clock in the morning, and then the third at five o'clock. In the evening after dark, the scuttles are closed, the lanterns lit, and the crew lines up on the port and starboard sides of the deck. At the roll of the drum, hats all come off and the helmsman recites a prayer. Immediately afterwards, the master-at-arms reads the list of punishments, the men carrying the lanterns raise them above their heads, the top men climb onto the rails and pull off the tarpaulin which covers the hammocks and begin distributing them. Each man takes his own and carries it off on his shoulder with the large white roll on which he will sleep and climbs down below on the battery ladder. The gun deck now serves as a dormitory after having served as drill hall, reading room, galley, and crew's mess. Total darkness makes the lantern's light reflected on the polished barrel of a cannon that much brighter; the silence is interrupted only by the men's breathing as their hammocks lean with the ship's roll. On deck, the men on watch have gone to doze in the dark corners, while remaining ready to act on any order. The staccato bell sounds the hours while the officer on the poop deck, his hands deep in the pockets of his coat, his collar turned up, starts a walk that must go on for four hours without stopping.

Life on board is not as monotonous as one might be tempted to think; it is interspersed with incidents which take on the importance of events, such as meeting a ship, saluting it, and sending a telegraph, or sighting a whale. You become interested in the distance travelled in the past day, in yesterday's, today's, and tomorrow's weather. You note the ship's position on a map as a point in the middle of the ocean and estimate the number of days to sail before arriving. No, nothing is monotonous at sea, not even the immense circle of which the ship occupies the centre and which she appears to carry along with her. Its size alone is eternally majestic; the waves, the colours, the sky, all the rest changes constantly. Within an unchanging frame, the painting is always different. The spectator, by comparison, has his own person, his ship, between him and the horizon. His eyes stop at every wave, from the one which lifts him up to the one which follows it, the one with a foamy crest farther on. From liquid valley to liquid valley, from crest to crest, you are carried toward infinity by successive intervals, to a much more human infinity than that of the firmament, which crushes man's smallness too brutally. You spend much time thinking while looking at these things, and your thoughts are rocked just as your body is. Sometimes your thoughts leap forward and suddenly reach the destination to which the ship is sailing; more often, they return to land at the point where the ship began to leave its wake. You dash into the future or to the past, carried by hope, by memory, by a date which was enough to bring you back to the beginning of life's journey.

The crossing continues. We are now cutting diagonally across the Gulf Stream, that great river of warm water which leaves the Gulf of Mexico through the Florida Canal, follows the coast of the United States, turns out to sea to bathe Nova Scotia and the south coast of Newfoundland, heads east, reaches Europe and warms its climate, and finally loses itself in the glacial Arctic Ocean, carrying with it clumps of seaweed, pieces of wrecked ships, tree trunks torn from the banks of the Amazon, which it deposits on the beaches of Iceland and Spitzbergen. South of the island of Newfoundland, this current meets the cold water arriving from the north, which is divided in two by the southern part of the island, spreads along the west coast and the east coast, and produces a condensation of vapour and creates thick fog.

With the movement of the Gulf Stream from the south, the temperature increases, the air becomes saturated with humidity, and you feel the same sensation as if you entered a steam room. The transformation is abrupt; a few hours ago we were in our normal state; now the thermometer has risen seven or eight degrees. We feel oppressed and short of breath, our nerves are tense, we feel a sort of languidness, almost suffocation, when suddenly the fog appears and envelopes the sea and the ship with a veil. In the trembling greyness where the reflection off the water looks like pale rays, you would think you were looking at the imprecise forms of fairies dancing above the moving waves. You then understand the poetry of the North, the misty lands of the elves, fairies, and Scandinavian *willis* and *norns*.[5] The southern sun is too bright; it gives only light and shadow and shows too crudely how things are. Dreaming is thinking about things that are not. The Oriental cannot dream; he can only stretch out his body, prostrated by the heat; he closes his heavy eyelids; his dreaming is *kief*, a state of drowsiness preceding sleep. A man of the north dreams while fully awake. Around the turn of a mountain path, in the midst of ruins, in front of bushes bordering the ravine and outlining the indistinct forms of the ruins, while the tall fir trees are moved by the breeze, their rustling can be heard and the torrent races along in its stone bed; the shiver of coolness stiffens your muscles and yet allows them to maintain their sensitivity. Only the traveller's vision is blurred; he thinks he sees and so he does. He thinks he hears and so does hear because his mind is full of strength and transforms the error committed by his senses into reality. His dreaming is active; if the Woman in White speaks to him, he sees Valhalla, that paradise of the brave where the defeated warrior is reborn to fight again, where eternally youthful Valkyries[6] refill the victors' cups with beer and mead. What a fine chapter a man of wit could write on the meteorology of passion!

The earth and sky make the man and his body, his intelligence, joys, enthusiasm, strength, and weakness, history and poetry. The great law of cause and effect holds true all the way to the most distant effect; when you reflect on it, you wonder about the exact meaning of human liberty. Could it not be that the *Faun*,[7] who for the Ancients was stronger than the most powerful gods, is the expression of that law and of our utter powerlessness to distinguish all existing mutual bonds? Because they are totally unconscious beings, minerals, plants, inferior animals, have no destiny inasmuch as all the events in which they are involved can or will be predicted; it is simply a matter of science. It would be the same for creatures of supreme intelligence; there would be no element of chance for them. Chance only exists for those who are as far from extreme intelligence as they are from extreme ignorance. Among them, chance varies according to the keenness of vision and the moral agility of the individual. For the ambitious, for example, destiny, the total number of causes producing a total number of inevitable consequences, can be likened to a terrifying railway train whose path at any given moment is determined by the previous moment. The partially sighted do not realize that they are following a track. They stop, the train passes and crushes them. The skilful carefully observe the huge machine as it approaches and roars; they take notice of its direction, and then step aside in safety. The cleverest start by seeking a safe place, await an opportune moment, and then jump on board the train, which carries them off and crushes the others. Of course, the slightest error is very serious; if their judgment is off, they will fall and be crushed, which is not a punishment but rather the consequence of their awkwardness. When one is not afraid of crushing one's neighbour, it is easy to succeed, and the talent of the truly ambitious is not as immense as you might suppose. It can almost invariably be summed up as a lot of flexibility and a very few preconceived notions.

The next day, the fog persists. We are now on the St-Pierre Bank, soon to arrive but yet unable to advance. The sea is covered with a cloak of lead and is smooth and ponderous; the ship hardly rises with the swell. The *Clorinde* tries to roll but cannot; she attempts to pitch, starts the forward motion, then stops, exhausted by her effort. On deck, you cannot see fifty metres. The whistle blows constantly, like a groan, to signal our position to any ships that might be close. The water condenses into large drops which run down the rigging or down the men's oilskins. Everything is dull and lifeless. Under the unhealthy influence of the warm, humid air, no one has any strength or energy. To be so near our destination – a few hours of clear sailing and we could be in the harbour at St-Pierre. We have been sailing for one full month; our

St-Pierre. A view of Île aux Chiens

fresh stores have been all used up, one by one, and our menu has become simplified: no more eggs, no more vegetables. The cattle that were brought from France, worn out by this long stay on board, yield very tough meat, and the chickens have become sea hens with meat so sinewy it resists the bite. Dried beans and lentils are followed by lentils and dried beans. After each meal, everyone morosely leaves the table and goes to a quiet corner or to the mess to recline on the cushions. You try in vain to read, to play, to talk, or to sleep. All you can do is try not to lose patience. Today, yesterday, and tomorrow all mean the same thing. For brief moments, the curtain seems to want to open, the horizon stretches a little, and hope is born anew. Soon, though, the opening closes again and you lapse back into torpor. In the evening, when we think we hear a ship's horn, we fire a cannon. The detonation is followed by a long rolling that echoes away in the distance, diminishes, increases in volume, comes closer, and then diminishes again until it is weakened by the foggy atmosphere and dies slowly away, leaving an even deeper silence all around us. We finally drop anchor.

Five days go by. On June third, at around two o'clock in the morning, I awake to the noise of men heaving at the capstan. Evidently, the weather has cleared and we are raising anchor. I dress hastily and go up on deck, where the night is clear and the sea as smooth as ice. In the distance, I see three dark land masses, the islands of St-Pierre, Miquelon, and Langlade. On the left, the two lights of Langlade, on the right, the Galantry lighthouse. We steam full speed ahead for fear of being trapped in the fog once again. We approach, and the coastline becomes more visible. We hug the cliffs of Miquelon, which rise steeply from the water and are bare of vegetation except for the top, where there is a carpet of deep green dwarf fir trees. Two or three patches of snow which turn pink at first light remind us that in this climate, winter is just ending in June. We steam on and pass between the island and an enormous rock called Grand Colombier, where a huge flock of gulls are resting or preparing to leave for the fishing banks. The harbour comes into view; in the distance lies the town and facing it, Île aux Chiens.[8] The screw slows down and stops, the chains slide with a roar through the hawsepipes, the anchors fall, and we are anchored two hundred metres from shore, opposite the dock at Cale Clément.

3

The Islands of St-Pierre
and Miquelon[1]

What a delightful feeling it is to arrive on a fine morning, peaceful, alone, and carefree in a new place for the first time. Nothing makes you think of any memory; you are not at all preoccupied. You are concerned with enjoying this feast of novelty for the eyes and the mind. Everything is unexpected; at every step the view is a new one. This is the pleasure that comes from discovering the unknown, without the slightest fear. In a way, you are born anew to these things, to these people of whom only yesterday you knew absolutely nothing and whom today you contemplate, examine, listen to, and assess with whatever your maturity has slowly acquired in strength, reason, taste, and judgment. You resemble a blind man to whom eyesight has just been suddenly restored. In this country, where you are merely passing through, no one knows you and you know no one. You are competing with no one and, though no one loves you, no one has any cause to attack you. You walk along with a bounce in your step, looking from side to side, in front of you and behind you. You listen to your inner voice and become so happy to have absolutely no enemies that the experience is quite moving. As for me, when I find myself in such circumstances, which thank God has happened to me more than once, if I chance to meet even a dog or a cat with an honest face, I always feel they are bidding me welcome, and I am overtaken with the urge to shake their paw in gratitude.

Such were a few of the reflections that crossed my mind when, after disembarking from the *Clorinde,* I set foot on the bottom

St-Pierre. The "barachois"

step of Clément's dock. The step was partly under water, covered with seaweed and quite slippery. I did not stand there for long but leapt onto the landing stage and had the immense joy, after thirty-one days at sea, of touching dry land. In St-Pierre, as soon as you take one step, you go either uphill or down. When you come ashore, you climb up a passage between two large, wooden buildings used as warehouses. You then follow the route de Gueydon, which comes to an end one hundred metres farther at Cap à l'Aigle and, in the other direction, leads into town following the shoreline. The oarsmen who took me ashore have left to go back on board, and I stop to sit on a rock; my number one me begins to show my number two me around this country with which he is as yet unfamiliar.

The view is superb. The *Clorinde,* having just arrived, is already getting washed and ready. On deck and in the rigging everyone is at his task. Boats come and go, sailors run along on the port and starboard beams. The frigate seems to bustle with activity like the traveller at the end of his journey who, with renewed enthusiasm, gathers the strength he needs to prepare his rest and to hasten the moment when he will be able to enjoy it. Behind me lies the rugged, arid mountain extending away as the hills recede behind one another and mounds conceal shelves until they gradually disappear in the middle distance at the Pointe de Galantry, crowned by its lighthouse. In the foreground, the town of St-Pierre, dark and flat, dominated by its high crucifix, climbs painfully up the hill and stretches nonetheless back to where I am standing, a line of houses spread out on either side of the route de Gueydon. Beyond Galantry, and separated from it by one of the harbour's entrances, lies Île aux Chiens, flat, covered with houses unevenly spaced around the church, which resembles an enormous barn, its steeple a tiny chimney. Leaving the Passe du Nord-Est, the Passe aux Flétans, and the Passe du Sud-Est[2] to themselves, are the other tiny islands: Île aux Vainqueurs, Île aux Pigeons, and Île Massacre;[3] farther away lies Île Verte,[4] and beyond, on the horizon, the coast of Newfoundland. It is the Feast of the Ascension. The sun is radiant and the harbour is full of ships, three-masted vessels from France, bound for the bank fishery with their crews of sailors, *graviers,*[5] and labourers; all have dories piled on their decks. There are sleek, oceangoing vessels, which take on freshly salted codfish and transport it to Europe as fast as possible to prevent the fish from changing colour and being damaged by the heat so that they obtain the best price on the market; but especially noticeable are the numerous schooners, with their pointed mizzen topgallant staysail masts removed and the mizzen stretched in a tight curve toward the bow, like American schooners. You also see steamships, entering and leaving port, rising and falling on the swell, most of them with their sails hung out to dry, lying limp along the mast. All these boats lean in the breeze, full of life and activity.

On land, in the town, on the hill, on the entire island, there is not a single tree to be seen. No matter what direction you look, your eye sees only rock protruding through the layer of grass and moss covering the mounds and crevasses, or houses built of wood, seldom painted, which the elements have turned some smoky grey, with their black, wooden-shingled roofs, where the eye searches in vain for the cheerful sight of red tile or blue slate. Everywhere there is rock. On the hilltops, enormous blocks have become loosened and hang on the edge, above the houses, until the day when the frost changes their centre of gravity and they are destined to come crashing down the slope and plunge into the sea, or join other similar boulders lying all around the shore of the island, haphazardly, piled together, large and small alike, forming a mass of fallen rock. Each salt storage house along the left side of the road known as route de Gueydon is connected to a wharf built on pilings and extending a considerable distance out into the water so that ships may come alongside to load or unload. The slope is so steep that the houses whose doors open onto the road are supported on the opposite side by thick pilings and often by solid beams serving as flying buttresses. Many have a flagpole and fly ship owners' flags or French colours in honour of the solemnity of the day. Along the right side of the road, on the way into St-Pierre, are private dwellings. They are all built of wood but are more carefully constructed, which shows that they are lived in; many of them have been coated with oil paint. You can recognize imported building materials from the United States, or more often from Canada, by the style of windows, without shutters or blinds, the porch often with a large window on each side and a balcony on the roof.

After less than twenty minutes' walk, you arrive in town. Occasionally, you have to cross a stream of yellowish, amber-coloured water, full of tannin from the peat, bubbling over the rocks, white with foam, in the hollows that descend from the hilltops. In many places, a narrow channel made of three planks nailed together has been built to divert a stream across the road, carrying the water to the end of a wharf where schooners can easily take it on board. Soon the road has become a street and the houses are closer together. All are made of wood, and most have a single storey; the boards on the outer walls overlap one another horizontally, and their roofs are sloped steeply in order to support the weight of the snow. Nearly all the houses have large double-sash windows, without shutters or slatted blinds, because there is so little light in winter. Behind the windows there are English-style green blinds, flowerpots with geraniums, souvenirs from Normandy, which no doubt require a considerable amount of care. The doorways are raised two or three steps above the level of the street and are protected by a porch from the *poudrin,* the fine, powdery snow, which can penetrate a house through the tiniest of openings.

Each house has a garden surrounded by a fence made of spruce branches, and containing soil transported from France,[6]

St-Pierre. Courthouse

St-Pierre. Quay de la Roncière

which is used to cultivate the few vegetables, which, during the summer, grow with the speed that is typical of cold climates. Turn to the right and you arrive at the square, which is decorated only by a fountain and where you will find the church, a very modest structure also built of wood, and the courthouse, St-Pierre's only stone building, with a zinc roof and high chimneys. Themis[7] has the best lodgings, which is fitting, since he probably has the highest revenues. The fable of "The Oyster and the Litigants"[8] is nothing new; it is only too true that our fortunes, which are amassed with such effort, preserved so courageously by such sacrifice and hardship, are taken almost entirely by the State, and most of the remainder by lawyers. Our children are the last to be considered and get only the leftover scraps. There is no point trying to avoid trial or flee discussion; we must let ourselves be fleeced, with the same resignation as angels, for resignation is the most important quality of both angels and men.

After the courthouse, you arrive at the Palais du gouvernement, the residence of the commander of the islands. The word *palais* or palace is exceedingly pompous for a one-storey building built of wood, next to the *gendarmerie* with its two barracks separated from the square by a wire fence and a double-stairway decorated with two oil-lit candelabra. All this lies in the shade of the French flag, the only shade on the island. It would be an exaggeration to state that it is majestic, but who is to say whether there is any point in having it any different? The interior is said to be quite comfortable, and if one were absolutely set on spending money, which would not be a bad thing since St-Pierre is one of our most prosperous colonies, it would be better spent on improving the harbour or constructing a dry dock to repair ships, like those which exist in Sydney and St John's, much to St-Pierre's disadvantage. The town is presently limited to the small *barachois*[9] which is inadequate for the numerous schooners that winter here and spend half the year in the harbour, caught in the ice as tightly packed against one another as codfish in a barrel.

The entrance to the harbour is opposite the government building's windows. The Quai de La Roncière is a vast open square bordered by shops. It has stone walls and large wooden landing stages that extend perpendicularly out into the harbour and thus increase its surface area. The square is the busiest part of St-Pierre and has a fountain with a cast-iron basin but no water, which is a precaution because of frost. There sailors gather, ship owners stroll

St-Pierre. Governor's residence

and carry on their affairs between the unhitched wagons, dories are pulled ashore, wood from Canada is piled up alongside casks of wine from France and cider from Normandy. Men work, children play after school, ships' boys fight among themselves, and dogs, made cheerful by the sunny weather, race around, much to the despair of a photographer – there are even photographers in St-Pierre – who patiently waits with his hand on the shutter until man's so-called best friend decides to exit the instrument's field of view.

Turning away from the harbour, you return to town. The streets are straight and very clean and intersect one another at right angles. There are many shops, especially in the vicinity of the harbour, and they all seem to sell the same objects. In the large windows, first of all you see fishing supplies: squid jiggers, hooks, cod jiggers, splitting knives, gloves. Then kitchen utensils, clothing, American-made boots, and liqueurs, creams, and elixirs in bottles of all shapes and sizes: round, stocky, tall, shaped like the Vendôme Monument,[10] or rather the Juillet Column,[11] the latter being more fashionable these days, political bottles, aristocratic bottles, ones with portraits of Messrs. Thiers[12] and Gambetta,[13] bottles of white, pink, red, and green liquid, bottles with bright-coloured labels, made of satin-like or gold paper with ornate illustrations akin to sculpture, to painting, to all the fine arts, for the use of those who are thirsty, and God knows they are many. The disease seems to be contagious because the first travelling companion we meet as we wander, looking all around with our head in the air, invites us to take a glass of beer in a café. So we go into a vast hall, where there is a theatre decorated with a portrait of Sarah Bernhardt,[14] and where, during the winter – a season when there is said to be plenty of fun to be had in St-Pierre – amateur actors give performances followed by a dance.

The upper part of the town is less aristocratic, and in the vicinity of the Crucifix, the streets are irregular and full of large stones, and the houses are more or less run-down, with no more than one storey. Each one is equipped with two ladders, one leading to the entrance, the other to the rooftop, so that at the slightest risk of fire, buckets of water can be carried up. There are countless children – a result, as they say, of the long winters and a diet of fish – and numerous hens, pigs, and dogs, all seemingly good friends, seeking their subsistence in garbage piles, sleeping in the sun; these wooden shacks are anything but opulent. On the other hand, from the foot of the Crucifix, there is a view of the harbour and ships. The islands form a chain extending to the left, while to the right are the mountains as far as Cap à l'Aigle, which reaches up vertically; in the distance, the peaks of Newfoundland occupy the horizon. The panorama is magnificent.

St-Pierre. Street [Rue Amiral Muselier]

St-Pierre and its neighbours, Great Miquelon and Petite Miquelon[15] connected by an isthmus, are the last remnants of France's possessions in North America. Under Louis XIV, we owned Canada. The intrepid Cavalier de la Salle, abandoned by all, with no other support but his own superhuman courage and tenacity, was murdered after having discovered the Ohio and the Mississippi rivers. At that time, we were masters of Louisiana to the south and to the north, Acadia, Cape Breton Island, Prince Edward Island, New Brunswick, Labrador, and a large part of Newfoundland. Today, we possess only these two islands and the right to station no more than fifty soldiers. St-Pierre is the most populated of the three islands and measures seven-and-one-half kilometres long by five kilometres wide, with a population of 6,300 inhabitants. The climate is harsh with long winters and short, foggy summers. Snow covers the ground from November until April. Often the north wind rises and blows the snow into a powder, which is picked up one minute and carried somewhere else the next and accumulates into enormous drifts. All traffic becomes impossible; the inhabitants are forced to remain indoors because anyone caught out in a storm would be blinded and would certainly perish, suffocated by the frozen particles.

Miquelon is wooded, but since it has no harbour, virtually the entire population lives on St-Pierre, where there are no trees. In 1885, the highest temperature recorded was 23°C in August and 18°C in January and in March. The average annual temperature is 5°C, which is the same climate as that of southern Sweden. St-Pierre is oval in shape and two-thirds of its surface is hilly, covered with small lakes or bogs with rocks visible everywhere. The other one-third is low lying, except toward Galantry, where the ground is higher, consisting of a vast marsh where it is impossible to walk without sinking up to your knees. There is, however, a recently constructed road to Galantry which goes around the *barachois*, across the Boulo bridge, climbs gradually, and, from a distance of one kilometre, offers a view of the town spread out along its entire length, with the roofs of the houses scarcely separated from the mountain behind, the *barachois*, and the masts of the schooners, and in the foreground, the *graves* or "fields" of cod drying on stones.

A *grave* is indeed very much like a wheat field, the only differences being that instead of the soil being worked in even rows of green stalks and wheat, there is a layer of stones the size of a

St-Pierre. Cod drying on rocks and on "bordelaises"

human head covered with split cod, flattened and drying, and the smell of the fields is replaced by the stink of fish. To complete the comparison, conical piles of dried cod stand here and there and one sees in the distance, like harvesters, numerous *graviers*, alternately stooping down and standing up, picking up the fish or arranging them on the stones. The cod are also hung from *bordelaises*, vertical racks where the tails of the fish are placed between two horizontal boards so that the rest of the fish is exposed to the air. Another method is to lay the fish on racks made of branches or boughs, with one end lifted up higher for better exposure to the sun. The fish are turned over repeatedly and, at the slightest sign of bad weather, are all collected, loaded onto stretchers, piled in stacks, then covered with fir tree branches and wrapped with tarred canvas. As soon as the weather improves, the stack is uncovered, the fish are spread out, and the work starts over again until the cod are satisfactorily dried and can be put in storage. Until then, no one is sure of the harvest. The fishing boats may manage to escape the perils of the sea, which is often particularly rough on the banks, and may survive without losing a dory in the fog, and avoid the danger of being struck and sent to the bottom by one of the fast steamers crossing from Europe to Halifax or Newfoundland, which ply these waters continually, and then be lucky enough to find fish in the first place, which for some unknown reason does not always happen. Yet all the best efforts, hopes, and expenses of so many poor people are no guarantee that once the cod finally has been caught, split, cleaned, and salted, it will be not be damaged by heat or spoiled before it is dried, before being subjected to the hazards of the return journey to St-Pierre. When all those risks have been avoided through courage, skill, hard labour, and luck, once the fish is safely in St-Pierre, there is still the risk that it could be spoiled in a matter of minutes by a slight shower of rain or a sudden fog, which is enough to turn the fish into a mush that could barely serve as fertilizer. Or the wind may increase or the sun may heat up intensely, causing the surface of the fish to dry too quickly and harden while the inside rots. If one only knew what it takes to produce a dish of codfish, which is hardly nibbled at in some comfortable dining room, compared to the misery, the hardship, the loss of life, the suffering of widows and orphans! Bah, there is no use in thinking about all that.

The population of St-Pierre is composed mainly of Normans and Bretons, except for a certain number of Basques, Americans, and English from Newfoundland and Cape Breton. French is spoken with a Channel accent. I have in my possession a printed notice written by an Englishman who speaks the language of St-Pierre admirably and writes it just as he speaks it – but I do not mean admirably. It is nearly impossible to read this poster. If, on the other hand, one reads it aloud, it is easy to understand but gives the impression of listening to a peasant from Normandy.

Produt pour vande à bord de la goilette. Saint Martin's pact
pataïb, shourave, foit, avoine, carat, bette, poinoit, lard, poit
à soupe, barlit epluche, 6 boite deuf, 175 peo est choson. Applique
à bord au capitaine.

Which means, "Produce for sale aboard the schooner *St-Martin*: potatoes, cabbages, hay, oats, carrots, beet, parsnips, lard, dried peas, hulled barley, six crates of eggs, 175 pairs of socks. Apply on board to Captain ..."

St-Pierre. Galantry lighthouse

The Galantry lighthouse is built on a rock located on the top of a hill and consists of a large square tower with the light above, and on the ground floor, the office, the flag room opposite the lighthouse keeper's quarters, and in the middle a flagpole used to signal ships that arrive. Near the water's edge, on a small mound pounded by the waves, is the steam whistle, which, during foggy weather, produces a long whistle at regular intervals, a hoarse horn that can be heard from a distance of eight miles, a signal to ships lost in the fog.

St-Pierre is the land of cod. Everything reflects cod: when you go for a walk, you see cod and unfortunately you smell them too. In the shops, utensils for the cod fishery, along the docks, salt for drying and salting the cod, on the wharves and the streets, fishermen. If two people are talking, the fragment of their conversation you overhear contains the word cod, always cod, nothing but cod. If you ask twenty people their opinion of St-Pierre, nineteen will be unanimous in declaring that the island is an abominable

St-Pierre. View from the road to Galantry

place, a sterile rock without a tree or a river, without the slightest object to please the eye. I am careful not to deny the veracity of these accusations, though I admit that I do not find the island so terrible. It possesses the virtue of being out of the ordinary, not featureless but rather original. I am sorry not to see it in winter; I have only seen it in the fog and actually find it has a certain charm. Each country must be judged in weather that is typical of it, Spain in summer, Russia in winter. Dare I say that fog is the weather which best suits the particular kind of beauty that St-Pierre holds?

I remember a walk I took one morning in order to explore the mountainous part of the island. From Clements' wharf, I began to climb the slope covered with stones that have fallen from higher up the mountain. Ten minutes later, I was on top of the first rise, on the shore of a small lake of yellow water. The fog had almost disappeared and no longer interfered with the walk; it formed a very light veil, which the wind carried here and there, passing in front of the neighbouring peaks and veiling their contours, then hiding them completely before letting them reappear

34

soon after. In the hollows of the rocks, a slender plant showed its head with its pale, timid flowers, each leaf, each petal dotted with thousands of droplets of water. I felt that intimate sensation so aptly conveyed by the English word "chill"[16] a pleasurable shiver which, in spite of warm clothing, grips and penetrates you completely, right through to the marrow, as modern writers would say. I was savouring that delicious impression of solitude and thinking of Minna and Brenda[17] on their rock in the Orkneys, looking out upon the waves. In the pools of water and marshes, I walked on pretty bouquets of *Sarracenia purpurea*[18] whose urn-shaped leaves are reputed to cure gout and rheumatism. I felt more fortunate than Jean-Jacques in his desert;[19] I had nothing to fear; I was alone. I followed the ridge of hills, came hurtling down the slopes, climbed back up another slope, jumping from one clump of grass to the next to avoid sinking into peat moss. Occasionally my path was obstructed by a lake that I had to walk around. At other times, it was a stand of birch, fir, or juniper, a Lilliputian forest, where the trees are levelled by the onshore wind to a height of about sixty centimetres, and grow so tightly together that the best thing to do is to walk boldly on top of them on the flexible carpet formed by their top branches until you step where the branches conceal a hole and you sink to your waist amid a bundle of stalks and roots from which you can hardly extricate yourself.[20] Fir trees here are a valuable resource, like grapevines; in the spring, when the sap is running, the young shoots are cut and steeped with molasses and allowed to ferment, producing spruce beer, a very inexpensive beverage that has the merit of being wholesome but is atrociously unpleasant to taste, in fact, like everything that is healthy.

I continued my walk, breaking off pieces of striped rock that the frost had shattered into sharp fragments, which accumulate beneath the large boulders. I discovered amethysts and veins of beautiful white quartz, so resistant that frost had no effect on it at all, and had left it intact on its own or in thick layers which show to what extent erosion has affected the surrounding rock. I then encountered another stand of fir, and after a considerable amount of gymnastics, managed to arrive at the other side, out of breath. I lay down to rest on the grass. From where I was, I could distinguish, through a clearing, Colombier and the mountains of Miquelon and Newfoundland. The sea was dotted with white, the sails of schooners leaving for the fishery; I deluded myself into thinking that this entire spectacle, which I was alone in admiring, was created entirely for me and had been awaiting my arrival. Then a breeze passed, the open curtain in the fog closed again. I stood up and shook off my vanity along with the twigs and bits of grass on my clothes and set on my way again. What a delightful sensation it is to feel free, with no one to stop the movement of your elbows or your thoughts, to gaze only upon the wilderness as it was a thousand years ago, before man deformed, spoiled, brutalized, and defiled it. What a joy to breath fresh air, to feel cooled by the wind that blows the clouds away and with them all cares, worries, and woes, to be conscious of the ineffable happiness of being as free of all hope, regret, and hatred as an infant, full of its mother's milk, warmly clothed, on its mother's knee, with pink cheeks, fists rolled up and eyes open to look at the angels flying above.

I climb down through the bushes to the route d'Iphigénie, near Anse de Savoyard; I pass in front of Robinson's, a lonely café and dance hall, without a tree or a flower, which looks as though it is brooding; in front stands the Iphigénie obelisk, a column in the middle of the round-about. The marble plaque at its base serves as a target for hunters, judging by all the buckshot marks; it was erected in memory of the sailors from the *Iphigénie* who built the road. I walk along the ponds that overlook the town and supply its water; they appear indigo-blue from a distance and a magnificent reddish-brown when you look through their water at the stones on the bottom. I follow the streets down from the Crucifix through St-Pierre along the route de Gueydon and arrive at the dock just in time to board a launch heading out to the ship. It has been a good day and I have earned the right to stretch out on the cushions in the mess and look out at the schooners that pass close by our stern as they enter or leave the harbour.

The island of Miquelon is separated from St-Pierre by a strait some three miles wide and is, in fact, made up of two islands, Langlade to the south and Grande Miquelon to the north, joined by an isthmus ten kilometres in length and three hundred metres wide at its narrowest point, and about three metres above sea-level, presenting one of the worst hazards to navigation in this area because it is invisible to ships and the current leads directly to it. Grande Miquelon is square in shape, its sides approximately facing the four points of the compass. The village of Miquelon is situated near the northernmost point, on the edge of a lake,

St-Pierre. Near Cap à l'Aigle

St-Pierre. Iphigénie Monument

and like the Grand Barachois to the south, offers no shelter to ships seeking refuge.

Langlade is practically uninhabited; the governor comes to holiday from time to time in a house that is almost always empty. Another house serves to shelter the *gendarmes* stationed there to maintain law and order, and there are a few farms where cattle are raised. From the boat, I noticed a few fishing shacks in some of the coves, between the cliffs. Both islands are mountainous and wooded, although trees are of very mediocre height, the ground covered with bushes and generally swampy. Overall, they are infinitely more pleasant than St-Pierre; the fact that they are abandoned can be blamed on the absence of any kind of sheltered port or harbour. Below the governor's country house, which overlooks the sea, flows a beautiful stream with clear though peaty water worthy of its name, la Belle Rivière, which winds its way through delightful grassland and continues inland, where it becomes difficult to follow for lack of a path or road. When fog prevails in St-Pierre, Langlade is usually bathed in lovely sunshine under a blue sky, as it was when I had the good fortune to make a short visit.

The isthmus is the most interesting part of the islands and stretches from north to south. On the Langlade side, stretching down from the foot of the mountains, lies a vast, mossy marshland, leading to sand dunes covered by a variety of grasses where the wind sweeps the bare sand over a large area. Near Miquelon, where the isthmus is wider, lies Grand Barachois pond, beyond which are the lower slopes of the mountains. Along the first one-third of its length, the isthmus is covered with shipwrecks. I counted seventeen on the west side and two large ones on the east. The overall impression is one of gloom. The wrecks are as recent as a few days, weeks, or months old, side by side in death, some with their hull intact, some with their masts and rigging, others with only their ribbing visible, nothing more than an even row of timbers jutting from the sand, creating an impression of profound horror. Fog has rusted the iron; rain has washed everything clean but for long reddish-black stains down the sides of the shipwrecks, like blood from a wound. In winter, when snow covers the ground, these white ghosts stand beside the dark water of the sea beneath a sombre sky. This sight appeared to me in summer, under brilliant sunlight that warmed the sand and caused the light mist to rise and be carried off with the breeze; I contemplated the strange appearance of the mountains resting on a base that seemed to be continuously moving. The pure blue sky was spangled with clouds low on the horizon, and the sea was calm. Small waves broke on the beach with a deep, murmuring growl and caressed the wrecks of the ships they had brought to destruction. Pale green in colour and cold, they were the same shade as the eyes of a wild animal. What an insult such tranquillity is, a mockery to all who have died so valiantly. Half-buried in the sand, debris of all kinds is scattered here and there, tinged with a shiny grey, masts, broken main yards, smashed barrels, pulleys, stays, broken dories, pieces of cork used to float nets, cooking utensils, heaps of dead vegetation, big and small. Long, cylindrical or corrugated, black or pallid-yellow seaweed roots with brown spots, drying or growing among the ruins. A few flattened sea urchins and white shellfish that turn to powder at the slightest blow lie scattered about, and in places a dried, reddish foam that you avoid stepping in. Occasionally, a seabird flies quickly across the isthmus, but nowhere is there a green plant or an animal to be seen, not even a humble sand fly, which in other countries runs, jumps, lives, and does its work on beaches. In the centre of the isthmus, to the west, there are three dark dots at some distance from the shore, which are probably remnants of sunken ships, partly submerged. The waves that wash over and then uncover them surround them with a border of white foam, and you would think they were human corpses, immobile and still resisting the sea. I stopped to observe the sight and, standing inside the skeleton of a ship with huge disjointed limbs, where men once worked, but without the soul that once gave it life, I felt as though my presence violated the silent peace of a vast ossuary.

A Little Geography and History[1]

Newfoundland is a large island with the Gulf of St Lawrence to the west and the Strait of Belle Isle separating it from Labrador and Canada. To the north, the east, and the south is the Atlantic Ocean. Its full length from Cape Ray to Cape Norman, exactly equal to its full width from Cape Spear to Cape Anguille, measures 511 kilometres. It has roughly the shape of a triangle whose peak looks toward Greenland. Countless bays and coves lined with mountains cut into its sides, especially from Cape St John to Cape Ray, and give it the characteristic appearance of shorelines with fjords, such as those in Norway, Scotland, and Ireland. We now know that a fjord is a place where a glacier once descended to the sea. Following a warming of the earth's climate and a sinking of the earth's surface, the ice disappeared, the sea flowed in to fill the space, and today, there is only a long, narrow, deep bay with steep sides often extending to a series of islands in the sea. To the southeast, where the capital city of St John's is located, the Avalon Peninsula is connected to the rest of the island by an isthmus whose width at the narrowest point does not exceed three kilometres. The interior is still practically unexplored. The Long Range Mountains, a low chain with the exception of several isolated peaks called *tolts*, extend northeast to southwest and determine the general direction of the rivers, lakes, and coastal indentations. As in all other low-lying countries, the waters on the same plateau sometimes run in different directions. As in the United States, where you can travel by water from the Atlantic to the Gulf of Mexico following the Mississippi river system from its source, it has been proven that in the Montana National Park, when two leaves fall from the same tree, one can be carried to the Pacific and the other to the Atlantic by the Missouri and the Mississippi rivers. In Newfoundland, lakes are most numerous in the east and the south, although the largest ones are found in the north and the west.

Geographical names are English and French; for the latter, it sometimes happens that their pronunciations or spellings are strangely distorted by the English language. It is difficult to recognize Cape Braha as *Cap Bréhat*, Boncer Bay as *La Roncière*, Baccalieu and Groais islands and Great Barrisway for *bacalao* (the Spanish word for cod), *Groix*, or *Grand Barachois*.[2] This latter name is common for places in this region and signifies for sailors from Normandy, a bay that is well enough protected to serve as shelter. Other names of lakes are Spanish or Micmac[3] Indian in origin and not as easy to pronounce, such as *Ahwachanjeech, Elnuchibeesh, Wachtapeesh,* and *Wagadigulsiboo Gospen,* for example. Such is the history of this country, where each person named whatever place he discovered or lived in: French sailors named the islands and bays in the north and south, the English those in the southeast, the native Indians, the rivers and lakes in the interior. Moreover, the transformation of names when they are borrowed from another language is a frequent occurrence in all of North America. In the United States, in the state of Iowa near Galena, there is a river called the Fever River, which was simply the Rivière de Febves or Fèves,[4] named by French explorers. When I was carrying out a topographical survey in northern Minnesota, I remember naming a lake after a Monsieur Maunoir, the likeable Secretary General of the Paris Geographical Society. The lake was one I had taken pains to choose among the most picturesque, in honour of its patron, but it had the misfortune of ending up being called Manure Lake! Which goes to show that you can encounter odd things everywhere, even in geography.

No matter what others may think, I love to look at the map of a country and the names that are written on it and read the history of the first inhabitants to live there. Their story is often a rather sad one, though very poetic, a story of struggle, hope, trouble, memories, and despair. On the map of Newfoundland, Fréhel, Groix, Belle-Isle, St-Lunaire are reminders of the poor Breton sailor preserving the memory of the land of his birth, when he saw resemblances more with his heart than his eyes and gave familiar names to the fog-shrouded capes of these wild islands. The banal names Grand Chat, Baie du lièvre, Baie au lapin[5] tell of the boredom of a long sailing trip where the tiniest incident, a hare escaping, the ship's cat sleeping on a coil of rope at the moment land is sighted, everything becomes an event. Come by Chance was named by a philosopher, White Bay for the icebergs that float there, Red Cape for its purple granite cliffs, Flower's Cove by a humble artist, Notre Dame, Trinity, and Conception Bays by pious believers, Bay Despair,[6] Isle aux Morts, Trepassey Bay,[7] all refer to the sailors' miseries. You can become lost in thought while reading a map, as you can while reading a calendar. What can be more agreeable on a winter's evening by the fire, when the snow is falling or the north wind blows outside, than reading the newspaper and being reminded of all those absent, far beyond the ocean, and those sleeping in the ground? They come to mind and reawaken when their names are evoked. They carry us back to our youth, while the eye looks on the dying flames without seeing them, the ear quivers at the crackling of the embers, and the soul smiles and takes flight toward the past.

In 1874, the population of Newfoundland was 159,033 inhabitants, of whom 95,000 lived on the Avalon Peninsula, an English colony. It has Responsible Government, made up of a Governor appointed by the Crown,[8] an Executive Council of seven members, a Legislative Council or Upper Chamber of fifteen members, and a Lower Chamber of thirty-one members elected by the population. The principal officers of the government are chosen among the latter by majority vote. The island is divided into six electoral districts comprising a certain portion of the coast, the total accounting for two-thirds of the circumference of Newfoundland. The remaining one-third is the French Shore, which extends from the west to the north, and the northeast to Cape St John is virtually excluded from the control of the Department of Colonial Affairs. The population is sedentary and is still quite sparse.

Newfoundland's climate is cold and, above all, very changeable. St John's is snow covered for seven months a year. During the summer months, an ocean current from the Arctic carries enormous icebergs along its east coast; when the current comes in contact with the Gulf Stream, which washes the south coast with its warm waters, it produces thick fog. Newfoundland is the land of fog. It does not penetrate more than fifteen or twenty leagues inland. But from Cape Ray to Cape St John, to the north, ships can sometimes be prevented from sailing for two full weeks.

The soil is poor. The rock, broken into pieces by frost, forms a thin layer of mineral debris, mixed with organic matter, on which mosses grow, saturated with water like a sponge. During the summer it forms a marsh and in winter a layer of ice, covered in snow. The Avalon Peninsula, Notre Dame Bay, and the interior of the island, however, have better soil. Although the trees do not grow as high as they do on the continent, vegetation is quite dense and is composed of fir, birch, ash, poplar, willow, maple, and alder.[9] There is no cedar, beech, elm, or oak, all of which are common in Canada. Forests are the home of caribou, black bear, and other fur-bearing animals, and seals and walrus are found on the coast. The coastal waters abound in an incredible variety of species of fish – cod, turbot, plaice, capelin, herring, and lobster – and trout are plentiful in the fresh water. However, wildlife is far less abundant here than in Canada,[10] and many kinds of animals that are common there, such as squirrel, mink, skunk, lynx,[11] and several types of fish, including pike, are non-existent in Newfoundland.[12] This observation would tend to prove that Newfoundland is not as new as its name would suggest and that it has been separated from Canada for a long time.

Because of France's particular situation in Newfoundland, continual discussions take place between our diplomats and the English, and, instead of becoming more serene, seem on the contrary to become more and more complicated. It is obvious that talks will conclude some day – everything in this world comes to an end eventually – what is not known is whether the solution, when it is arrived at, will be advantageous or not for France. When you look at the question closely, you quickly realize that it is intricately woven into a great number of subsidiary questions which must be dealt with first. I myself do not possess the qualifications to participate in such a serious and complicated debate. I am merely expressing my conviction that it cannot continue much longer,[13] and I would like to present to the reader a historical summary of the various phases which have led to the present situation.

The island of Newfoundland was discovered on June 24th 1497, by John Cabot, a Venetian explorer in the service of King Henry VII of England. He was captain of the *Matthew* out of Bristol and took possession of the island on behalf of the King. It is possible that it had been visited and perhaps even colonized as early as 1002 by Icelanders and Scandinavians. French Basques claim to have made numerous voyages there during the four-teenth century. As in all matters pertaining to the ownership of the discovery, science is unable to fully separate fact from legend. The island was then visited by Jacques Cartier from St-Malo and then by Champlain. In 1525, Verazzani was sent by François I, with special instructions to take possession of the island. In fact from 1504, it was known how abundant cod were in these waters, and fishermen from Normandy and Brittany came regularly to fish here. In 1517, the number of ships from various nations reached fifty; in 1578, France alone sent 150 ships to the banks, Spain 125 or 130, Portugal 50, and England from 30 to 50. In 1583, Sir Humphrey Gilbert took possession of the island for Queen Elizabeth, and did so quite effectively, for he brought 250 colonists.[14] In 1604, French fishermen founded Plaisance [Placentia] on the south coast, their first sedentary establishment, and during the entire seventeenth century, on the southeast coast, they fought constantly for possession of the island with the English, who were established in St John's.[15]

Under the administration of Colbert,[16] the French Navy was undefeated in the Atlantic and in the waters near the Gulf of St Lawrence; its warships patrolled the seas between France and Canada, constantly coming and going. French fishermen had de facto exclusive fishing rights on the north, south, and west coasts of Newfoundland, where the English, limited to the Avalon Peninsula, were careful not to interfere with them. But during the last years of Louis XIV's reign, the situation changed dramatically. France was defeated in Germany, Italy, and the Netherlands and lost its navy. The Treaty of Utrecht (April 11th, 1713), which took back land in Hudson Bay and Acadia and gave it to England, determined Newfoundland's fate. Henceforth the island belonged to England; as for France, it kept the right to fish and dry fish along the entire coast from Cape Bonavista on the east coast, north to Pointe Riche on the west coast. We maintained Cape Breton Island, however, where the town of Louisbourg was built; the Grand Bank fishery remained open to us and has been ever since. These rights were confirmed by the Treaty of Aix-la-Chapelle in 1741[17] and by the Treaty of Paris (February 10th, 1763), by which we lost all of Canada, Acadia, Cape Breton, and all the islands of the Gulf of St Lawrence. As sole consolation, we were granted the two tiny islands of St-Pierre and Miquelon. From 1713 to 1760, the fishery was the livelihood of 16,000 Frenchmen.

In 1778, the English seized the islands of St-Pierre and Miquelon and destroyed all buildings, forcing the inhabitants to seek refuge in France. The Treaty of Versailles (September 3rd, 1783) gave back the islands to France, and to avoid further trouble in Newfoundland, it was agreed that France would abandon its rights to the coast between Cape Bonavista and Cape St John in exchange for equivalent rights on the west coast between Point Riche and Cape Ray.[18] At that time, France was in a good position to ensure that its rights were clearly defined, for the United States had just freed itself from the yoke of England and our navy, reorganized by Choiseul[19] and under the command of Orvilliers,[20] d'Estaing,[21] Grasse,[22] and Bailli de Saffren,[23] had almost recovered its former supremacy. Unfortunately, de Vergennes[24] was unable to benefit from such an advantage.

In spite of the fact that St-Pierre and Miquelon was taken by the English on May 14th, 1789,[25] the 1783 declaration, which restored the tiny islands to France, was renewed by the Treaty of Amiens in 1802. After St-Pierre and Miquelon was lost once again in 1803, according to the terms of the Treaty of Paris (June 30th, 1814), it was ceded back to France. At long last, a century after the Treaty of Utrecht was signed, its terms were confirmed.

However, for various reasons, matters soon became more complicated. In the first instance, the population of Newfoundland was rapidly increasing. With the marvellous skill at colonization that the Anglo-Saxon race possesses, every year swarms of families from the Avalon Peninsula left St John's to settle in places where life would be easier for them, that is to say along the coast, because the sea is an open road for communication and supplies. The interior of Newfoundland is a wilderness, without a single road or trail, and is virtually unexplored. The soil is poor; agriculture is precarious, if not impossible, except in a small number of places, which were the first ones to be colonized. The fishery, on the other hand, offers bountiful resources that are immediately available. Our fishermen, however, forced to make the dangerous crossing of the Atlantic to reach their fishing grounds, have very high expenses, unlike the Newfoundland schooners, which follow the cod without being confined to a fixed residence, and which are, in many ways, ready for action, so to speak, with fewer expenditures and less time lost. Their crews "fish" for seals, which is only possible during the winter; during the summer, they are practically idle and content with a small salary in addition to whatever profit they make. They are certain to sell the small cod, which are consumed locally, and do not concern themselves with the damage caused by their narrow mesh nets. Our fishermen, meanwhile, even supposing that they were tempted to use such illegal equipment, would find no market for fish that are considered too small to sell or, at least, would only be able to sell them at a price far below what would be required to recover their costs.

This question involves another factor which must not go unnoticed. I am referring to that special tenacity which characterizes the English, which enables them to benefit from everything that they can turn to their advantage: their strength if they happen to be strong, their weakness if they happen to be weak, even the pity which they inspire in some cases, which enables them to encroach slowly and steadily upon their neighbours. Always ready to advance, they never retreat unless obliged to do so by force, and then they come back obstinately as soon as that force relaxes even slightly. How can you force those unfortunate mortals who are in the depths of misery to respect treaties? Can you prevent the hungry from feeding their families with the bountiful manna from the sea in front of their very dwellings? So it happened that the English settled first in Cape Ray, then St George's Bay, Codroy, Bay of Islands, Bonne Bay, White Bay. In the name of compassion, it is tolerated for one year, then the following year what was tolerated comes to be perceived as a right. Moreover, in most cases, the places were vacant to begin with. When a French cod fisherman complains, the administration first studies the matter; by then three years have passed. The fourth year, a commissioner is sent to investigate and attempts to conclude an arrangement. By then five years have gone by, or perhaps even six, because it is to the advantage of the English to have the matter drag on.

During this time, the sedentary population has been increasing in size, working relentlessly, clearing the land, making roads, building houses, while during the same time, due to weariness or to other causes related to the overall economy, France's fishing boats have decreased in number.

Finally, after prolonged discussions, when it has become impossible to delay any longer and always after having realized a profit, sometimes small but more often large, London ends up reaching an understanding with Paris and an agreement is signed. This brings about another change. According to the relations that have existed since 1855 between England and its colony, this agreement, which was so difficult to obtain, must receive approval from the House of Assembly of Newfoundland in order to be valid, but such approval is denied for reasons that are quite understandable. Therefore, nothing more is done. The sedentary population has continued to increase in the area in question, and our ships have continued to dwindle in number. Certain circumstances, such as the Crimean War, for example, helped bring about a temporary solution, which of course was in favour of our neighbours since it was precisely in 1855 that Newfoundland declared independence, without France even appearing to grasp the significance of such an event.[26] New complaints are made on behalf of the few fishing boats which are persevering and stubborn enough not to give up and seek more profitable returns on their investment elsewhere. France, ever dignified, decides not to recognize the existence of the House of Assembly of Newfoundland and deems it appropriate to deal only with London, and so the entire matter starts over again. A commission is set up, discussions are undertaken, an agreement continues to be as elusive as ever and is reached only after protracted efforts. Finally, by dint of good will and thanks to a number of small sacrifices which, when added to those already conceded previously and which have been maintained religiously by the English, constitute very sizeable sacrifices, a decision is made to draft an agreement. The document is sent via London to St John's, where in testimony to a remarkable singleness of purpose, the Newfoundland Assembly promptly and unanimously refuses to approve it. England protests and registers its distress and leaves us free to proceed as before. On the other hand, if by chance – and the risk is considerable – one of our fishermen contravenes one letter of the Treaty of Utrecht, if a Frenchman has the misfortune of erecting a construction that has the slightest appearance of permanency, the protest from the English side is immediate, and the delinquents are abruptly called back into line. All of this would be rather amusing if it were not so sad.

In order to show that this is not an exaggeration, one has only to mention the principal attempts at a solution that have been made since the Treaty of Paris in 1814.

In 1844, there was an understanding between Lieutenant Commander Fabvre and Mr Thomas, President of the St John's Chamber of Commerce and later with Sir Anthony Perrier, but no further action was taken.

In 1851, Assistant Commissioner de Bon and Sir Anthony Perrier were unable to arrive at an agreement.

In 1856, Commander Pigeard went to London and concluded an agreement on January 14th, which was ratified on January 23rd, 1857, according to which the French obtained exclusive fishing rights and use of the coast from Cape St John to Quirpon Island and from there to Cape Norman; the same rights on the west coast in the five harbours of Port-au-Choix, Petit-Havre or Petit-Port, Port-à-Port [*sic*], Red Island, and Codroy Island; a shared right to fish with the English from Cape Norman to Cape Ray; exclusive rights along the shore from Cape Norman to Point Rock; the right to purchase capelin and herring freely and without interference along the south coast; and the right to fish all inland rivers, to fish in common with the English in Labrador and the Strait of Belle Isle, to use and guard English and French fishing settlements in winter and summer, and to cut wood for ship repairs and fishing settlements. As soon as the contents of the agreement were made known in St John's, there was an uprising of the population. The English flag was carried around the streets of the city attached to a horse's tail, the governor was insulted, the House of Assembly of Newfoundland refused to ratify it, and England sent its deepest apology to France. It is worth remarking that the English started to show signs of becoming conciliatory precisely when Newfoundland had become independent.[27] It was obvious to anyone with the slightest awareness of the country's affairs that the House of Assembly would never grant its approval to any sort of compromise.

On May 1st, 1859, an international commission was held during which France was represented by Captain Montaigne de Chauvanne and Monsieur de Gobineau, Secretary to the Ambassador, and on the English side, by Mr Kent, Colonial Secretary for Newfoundland, and Captain Dunlop of the *Tartar*.[28] Relations were most cordial, and a large number of English and French witnesses were heard. Proceedings lasted until August 27th aboard the *Gassendi*; then the conclusions were drafted. Negotiations continued in Paris, until March 1860, between Captains Dunlop and Montaigne, and new conclusions were drafted. The population of St John's showed its hostility, and the matter dragged on during 1860 and 1861, and when all was said and done, nothing had changed; the status quo remained.

After 1858, difficulties of another sort were added to those relating to the fishery and further complicated an already complex situation. That year, mineral deposits were discovered in the interior of Newfoundland and could only be shipped out through that part of the shore which was reserved for the use of the French. The English were forbidden to erect any permanent structure on any portion of the French Shore. In 1866, the British government charged Mr Musgrave, the Governor of Newfoundland, with the responsibility of bringing the Newfoundland Assembly to a resolution which could serve as a basis for renewed discussions. The resolution, dated April 8th, 1867, was submitted to Commander de Lapelin. The document consists of a single declaration on the part of the English that they possess all rights, as though the Treaty of Utrecht, and those which followed up to the Treaty of Paris in 1814, had never existed. It was difficult to arrive at an agreement. Fortunately, or unfortunately, depending upon your point of view, the matter resolved itself by the fact that the mines which threatened to become a new source of conflict proved to be unprofitable and were quickly abandoned.

In 1884, another attempt was made to conclude an agreement, but like the previous ones, it led only to the Newfoundland House of Assembly rejecting it once again.

The history of these negotiations has been presented in minute detail and with remarkable precision by the anonymous author of an article published in the *Revue des Deux Mondes* in 1874, titled "Terre-Neuve et les Traités." No major change has occurred since that time. The author argued in favour of England buying back France's rights and expressed the hope that Newfoundland would soon become part of the Dominion of Canada, which would hasten a mutually satisfactory solution, thanks to the good relations we have maintained with that country. Such a union, which was expected within a maximum of two years, has still not taken place.[29] I do not know whether it is still a probability, for in politics, good relations bear far less weight than self-interest. In point of fact, the situation is awkward on both sides. No agreement can be reached as long as the terms of an agreement signed in 1713, then revisited in 1741, 1763, 1783, 1802, 1814, 1844, 1851, 1856, 1859, 1867, and 1884, continue to be disregarded. These are nothing but diplomatic games. Nowadays, diplomacy is able to put things in check momentarily or force a hasty solution but never to modify it, because it is the inevitable consequence of a combination of facts. Words, whether spoken or written, cannot possibly prevail in such cases. No matter what the treaty is, when it is in a country's interest to violate it, when there is a vital interest in violating it, when it is forced to do so under pain of death, it always finds the chink in the diplomatic armour. In this case, its task is an easy one. The Treaty of Utrecht was as bad for Newfoundland as for the dispute over lands between France and Brazil in the Amazon Basin; this matter has been at exactly the same stage for 174 years. The Treaty of Versailles is equally imprecise. If we rise above these subtleties and attempt to find the spirit which animates instead of the letter which kills, we see that these treaties endeavour to establish an absurd state of affairs, an unstable political equilibrium doomed to fail, that they set out to prevent a people, the Anglo-Saxon race, from developing. A Scottish fisherman, tall and strong and eager to work, starving to death in his own country, accompanied by his wife and a string of children which gets longer each year, arrives in a lonely bay, builds himself a hut and moves in and, by dint of hard work, decides to earn himself a place in the sun, or rather in the snow. One morning, hungry, suffering from the cold and the thousand miseries of life in the wilderness, the storms on a sea that is his only source of food, he is informed that he must move away and abandon the place, not to other starving people, in which case the battle would be fair, based on the principle of an eye for eye and a tooth for a tooth and may the fittest and strongest survive and find food, no, but to the mosquitoes, in the name of an arrangement that had been concluded in Holland by people who have been dead for a hundred and fifty years. The Scotsman shrugs his shoulders, and who can blame him? Newfoundlanders want to live in Newfoundland; on its shores, their population is increasing as fast as ours is decreasing.[30]

Let us take a fair look at the respective situations of the three parties concerned. England has the responsibility of the naval division while Newfoundland is uncomfortable with the existence of poorly defined but legitimate rights. A young colony has better things to do than use up all its energy in endless conflicts and has everything to lose by stirring up animosity. Recent discussions with the United States must give it cause to reflect. Since the fishery is carried out in the most harmful way possible, it is in Newfoundland's interest to actually regulate it, because this source of wealth will soon be exhausted forever. France spends huge sums of money to maintain its naval division and pay subsidies to the ship owners for the cod fishery. For every boat that is sent from France to catch and dry cod in Newfoundland or in St-Pierre, the owner receives a subsidy of fifty francs per man if there are a minimum of fifty men and if the ship is at least 158 tons or at least twenty men if the tonnage is below 158. Ships that are able to salt the fish on board receive an additional thirty francs per man, with no minimum number required.

Under these conditions, why not arrive at a compromise? It is said that a bad deal is always better than a good trial. Let us attempt to sell our rights, whatever they may be. Newfoundland and England would not lose by buying them, and by ridding ourselves of them, we would be doing well. If the fishery became free around the island or even limited to the banks, it would suffice for the training of our sailors, since that is the reason – however questionable it is – that is always put forward when talking about Newfoundland. We could thus concentrate all our efforts on St-Pierre and Miquelon, the two islands which in spite of their tiny size are our most prosperous colony and would be even more prosperous if our ships found a safer haven than the present harbour and shipyard. This would not require a huge expenditure; profits could be made on foreign vessels and the investment would be worthwhile. Everything would seem to favour such an amiable settlement. Fishing grounds around the island are selected by lottery for a period of five years; and on each occasion, the number of boats decreases; the cod are actually growing scarcer each year, so there are fears that the next five-year lottery that decides who fishes where may actually be abandoned.

Will the parties concerned embark down a road paved with mutual concessions and a fair and intelligent policy? Will the solution be one that is recommended and hoped for by the majority of competent people who have an interest in this matter? As the Spanish say, *Quien sabe?*

5

Bonne Bay

Our frigate left St-Pierre in beautiful weather, with every man at his station, the captain on the poop deck, nearly all the crew on the capstan, half on deck, the other half on the gun deck. The men are four or five abreast, pushing on long wooden poles, heaving in time with the bugle and drum. We are setting sail for Newfoundland, for a new land at last, since St-Pierre is still a little part of France.

We glide past the rock of Grand-Colombier, which is the capital, the refuge, the roost of the puffin, a black and white seabird the size of a pigeon and resembling a duck that has a beak like a parrot and burrows into the ground like a rabbit. We sail past Langlade, along the isthmus and the west coast of Miquelon, and head northwest toward Cape Ray. We are soon out of sight of land and must wait until the next day to see it again in the distance, shrouded in the grey morning fog. That afternoon, we round Cape Anguille and Red Island and then sail close to the coast, where a vast panorama stretches before us, the details of which can be made out distinctly through the telescope. In places, the cliffs have been sheared off, leaving layers of sandstone or schist strangely twisted and dipping to the south, covered by greenish rock, giving the impression of a geological diagram. Here and there, a rock fall caused by frost and coastal ice has left piles of debris of some thousands of cubic metres in size, with huge boulders scattered throughout, partially submerged in the ocean. Tall mountains cast their slate-blue shadows on neighbouring slopes or on the greenish waves. The sky is a bright blue, streaked with

Newfoundland (West Coast). Shoal Brook Falls, Bonne Bay

white, as in all cold climates, and the nuances of colour are very subtle; although the scene is full of majesty, the general impression is one of gentleness and harmony. However, as you approach, the foreground is spoiled by the garish yellow of the young birch trees. I did see Bonne Bay again in the autumn when the leaves were red and much more pleasing to the eye; but now it is early spring and a few patches of snow still cover the ground in places. Inland, the forest stretches away endlessly. The day is waning, the sun casting its last rays, and the colours of the mountains soften even more as the shadows of night envelop the earth.

In the morning we arrive at Bonne Bay, rounding the southern point. Bonne Bay is a fjord divided into two main arms, East Arm, with Savage Cove, and South Arm. Near the coast, the water is extremely deep; we sail so near the cliffs that to see the tops of the mountains from the poop deck, you have to raise your head.

This country was shaped by glaciers, and not a single ledge or platform breaks the vertical rise of the cliffs that border on the sea. Once you have entered the bay, some of the slopes are less steep and more like layered hills, part of which were formed by an accumulation of fallen rock, carried downstream by rivers and deposited around the estuary.

On the right is the village of Bonne Bay.[1] Most of the houses are built along the water's edge, with an adjacent wharf extending out on pilings, made of fir tree trunks side by side, so that schooners can tie up to unload the cargo of cod. Others are farther away and have their façades painted white and wooden shingle roofs. They are surrounded by recently cleared fields enclosed by fences; the tilled, black soil still has whitened roots of the trees that formerly occupied the site. In the middle of a stand of fir, the sturdy steeple of a wooden church rises upward, and cattle graze in the large open fields nearby. Smoke drifts up from the chimneys of some of the houses, and you can hear roosters crowing. Nature is awakening and so are the inhabitants; women come out to look at our ship. The *Clorinde* sails slowly in and turns a little; then the view of the sea disappears and the bay opens up farther inland. The mountains are closer together and, with a carpet of green vegetation covering the slopes and concealing all details, you can barely distinguish their shapes. The houses are spread along the shore, and when we pass the opening of a valley, we see at a distance of some ten miles the beginning of an immense plateau of yellowish-red rock without any vegetation, so huge that it towers above us and we feel crushed by it.[2] There are gaps on the flanks of the plateau continuing up to the ridges on top of the two uniform slopes, where enormous boulders have rolled down from higher up and are scattered about like so many grains of sand in the dust. We are told that this plateau extends into the interior of the island and is separated from the sea by a belt of forest that makes access to it quite difficult. If the polar climate ever warms enough to melt the ice that covers the ground in Greenland, it will probably look much like this. Closer to us, the black rocks are stained with white and ochre-red from dried lichen and coloured green by ferns that grow in rows in the cracks of the rock. No matter how vast the landscape may seem, instead of fear, it inspires admiration. It is so uniformly huge that it is uplifting for anyone who contemplates it. When the sun momentarily breaks through the clouds that have gradually covered the sky, its rays

draw sparkling lines on the dark surface of the bay. The frigate
drops anchor at the bottom of South Arm, at the foot of a strange,
thumb-shaped peak, and as soon as the boats are ready, we
promptly go ashore.

We head toward a narrow valley that is a continuation of the
fjord enclosed between two mountains. The seafloor rises up suddenly and appears so near the surface that, even in a dory, we
barely have enough water to float. The movement of the oars
frightens a number of plaice, which swim off and huddle against
the mud and rocks, where they lie still, like flocks of sparrows in
a wheat field that fly up when disturbed by a walker, only to alight
a little farther on. A moment later, the dory runs aground and the
sailors carry us ashore to prevent us from getting wet. We cross a
narrow pebble beach and enter the wood. We are surrounded by
vegetation, and a dome of greenery forms a canopy made of
birch, fir, wild cherry, and alder over our heads. The ground is lit-
tered with irises, wide-leafed fennel, violets as pale as those from
Parma, and bouquets of ferns, with the extremity of each leaf
rolled up into a "fiddle-head," which, because of the thin stalk
and bushy head, look like the ostrich feathers on the coat of arms
of the Prince of Wales. The ground exudes a particular smell, the
penetrating odour of the wild. During this cold-climate spring,
the snow has hardly melted, and the sun's first rays have hardly
broken through the clouds; each plant wants to take advantage of
the short summer that is beginning and hurries to turn green and
to bloom. A small river flows down the valley and has carried with
it sediments fallen from higher up the slopes; its slow current is
enough to transport them to the mouth and to deposit them in
the deep waters of the fjord, where they have accumulated and
formed a steep, underwater bank where the *Clorinde* went aground
last year. After spending so long cramped in tiny cabins, we enjoy
the pleasure of being able to move our limbs and feel an incredible urge to walk, run, jump, and move our legs and arms. Unfortunately, it is difficult to penetrate the trees and bushes that
have been cut down and that lie intertwined and impenetrable;
everywhere you tread, your feet sink into the wet moss. We have
to stop, content to breathe in the smell of the forest, and then
turn back.

On the left of the mouth of the river there is a cabin built of
planks. One of our group stops nearby to sketch the scene.[3]
While he is drawing, seated on a tree trunk, a small boy of about

five who was watching us, half-hidden behind the window, observing our movements, gets up his nerve. He comes out, stops for a moment, thinks it over again, is held back by his shyness, drawn forward by curiosity, starting and stopping alternatively, until finally he comes close to the artist and stands still with his hands behind his back, examining the pencil marks with profound attention. His clear victory makes his older sister feel brave enough to come in turn, carrying a younger brother in her arms. Then comes the mother with an infant and an older brother who was hoeing in a nearby field and, finally, the father. All these calm, peaceful people, standing in a group, savouring their admiration, carefully contemplate the progress of the drawing, occasionally exchanging a word or two or a brief sentence. We make acquaintance, chat, and after the usual questions about how severe the winter was, how the cod fishing is, was our crossing pleasant, the man explains to us that they have lived there for four years; he built his own cabin and cleared a part of the forest; the winters are long, but one survives, makes a living, and enjoys good health. After all, is that not what wisdom is? I do not understand why, with the exception of the sick and infirm, there should be a single miserable person in the world. Every man has more than the right, he has an obligation to create a family, have children, and be happy. When he is young and strong and feels that society around him is going to crush him and that poverty is near, unless he is a fool or a coward – in which case he deserves no pity – he has only to gird his loins and walk straight ahead until he reaches the first patch of uncultivated and uninhabited land. Thank God, there is no shortage of it, to the north, the south, the east, or the west; it can be found everywhere, without going very far. Once he arrives, he immediately starts his battle against nature, that noble struggle from which man, if he so desires, always comes out victorious. If he works – I am speaking not in theory but from experience – in return for his labour, the earth that he made fertile will perhaps bring him riches, but certainly independence, health, and happiness.

The northern shore of South Arm is so steep that it is nearly impossible to scale, even on all fours. The vegetation starts at the water's edge; on the opposite side a path follows the shore to the village of Bonne Bay. Along nearly all of its length, but especially on the side nearest the water, the houses are spaced out with English irregularity, for in a country like this, where land is not expensive, they do not wish to rub elbows with neighbours. The path is wide and mostly bordered with fields surrounded by "fences," rough-hewn fir tree trunks characteristic of North America. They are built without a single nail or piece of iron; all that is needed is an axe and a forest nearby. Extremely diverse in design, they are very often a zigzag, forming a series of alternating open and closed angles; occasionally, they are straight, with the fir trees laid horizontally, supported on each end by two vertical posts joined together at the top by a crosspiece. They are not easy to describe; a drawing would give a better idea of the different varieties; in any event, they are quite picturesque. When an American, whether in the United States, Canada, or Newfoundland, wants to make a field, he cuts down trees in the forest to a convenient

Newfoundland (West Coast). Shoal Brook Bridge, Bonne Bay

height for his axe – that is to say, about one metre from the ground – leaves the stumps to serve as posts, burns the brush and the branches on the spot, and then abandons everything. While the cows graze on the grass, the tree stumps slowly rot in the sun, the rain, and the frost. After a time, they turn silver grey and lose all their strength, so that a good blow with a sledgehammer is enough to knock them over. Then the soil can be tilled, since the roots left underground are no longer an obstacle and have disappeared without any work.

The path crosses two rivers, or rather two large streams where wooden bridges are built on four-sided pilings made of fir logs stacked on one another. The resulting construction is solid, lightweight, inexpensive, and quite pleasing to the eye. Shoal Brook has a bridge like this, where the water flows over a bed of manganese rock, a sort of serpentine with a surface so glossy that you would think it had been oiled. About one hundred metres upstream, you find yourself facing a waterfall hidden from view by a headland. The water falls from a height of some eight or ten metres, straight down over red schist, forming a white layer of froth; the sunlight glistens on the drops of moisture, which rise in clouds and produce a rainbow. The branches make a green arch over the cascade and pool below, where trout follow one another in the limpid, icy water. It resembles the set of a comic opera.

We walk on, following the slopes of the path up and down, where wild raspberry flowers grow on either side, and admire the constantly changing panorama before our eyes. Just before reaching the village, we meet the minister, a man who is still young and probably delighted to find someone to talk to, and we make acquaintance immediately after a cordial handshake. He places himself at our disposal and begins by leading us to his house, where he introduces us to his wife. We sit for a moment in the parlour of his "humble parsonage," as he calls it, decorated with the organ, an indispensable furnishing in all English homes and used to accompany Sunday hymns. An open door reveals a study full of books. The rooms have low ceilings and are rather cluttered. Although the temperature outside is quite cold, it is actually excessively hot; when you come inside, you feel as if you are entering a steam room. The minister tells us about the life he leads in Bonne Bay and his occupations: in summer, he hardly leaves his house because it is practically impossible to travel inland. In winter, the ice forms along the seashore and accumulates to a thickness of four or five feet, and it is covered by snow; then he is able to visit his parishioners. Following the coast on snow-shoes, he makes trips of fifty or sixty miles as far as Ingornachoix Bay[4] to the north and the Bay of Islands to the south, sleeping in fishing huts on the ice, rolled in a blanket. He also travels by dogsled, and just as he is telling us this, a large Newfoundland dog lying on the rug, seeming to understand that he is being talked about, gets up, stretches, and goes to lay his black snout on his master's lap.

Newfoundland dogs are numerous in Newfoundland, which is logical, but they are seldom purebreds, at least on the French Shore, where they are crossed with Labrador retrievers. The finest I saw were in fact in St-Pierre: completely black, with an oc-casional white steak, a crop on the chest, the inside of the mouth black, the tail bushy, the paws webbed, of medium height, much smaller than a Pyrenees. Their fur, long and curly and resembling karakul or Persian lamb, is used to make hats; it is so thick and oily that even after a bath the skin remains perfectly dry. The Labrador retriever is smaller and has a pointed snout, a long tail, and short fur. In winter, they are attached to sleds; in summer, they live off their own meagre resources. They can be seen roam-ing along the shore starving, with nothing to eat but cods' heads and capelin that have washed ashore and dried in the sun, which they only pick at and which give them skin disease. They are quite friendly to humans but very aggressive toward one another, and our poor friend Lancelot, a basset hound that travelled from France with us, would not dare leave our side during our walks for fear of attack. This inhospitable welcome contributed to his melancholy and made it seem as though he were searching for his handkerchief to wipe his tears as he looked sadly around him. Newfoundland dogs are rather intelligent; they like to swim and carry out rescues in the water. They can save anything that floats, pieces of wood, algae, froth, and even a man, provided he does not move, because if he does, the dog will put his heavy paw on the head of the poor unfortunate, flailing about in the water, will stay within reach and keep him underwater until he drowns, and then seize him in his jaws and bring him ashore. If I am ever in danger of drowning, God preserve me, I hope with all my heart there is no Newfoundland dog nearby! I once knew two dogs that were raised together as friends, then became mates and made an ideal couple. One day, we threw a piece of wood in the water, and they both jumped in together. The female arrived first and grabbed the object in her teeth; the male arrived then and put his paw on her head. We had the utmost difficulty preventing him from drowning her. It is not easy for me to destroy an illusion that is dear to many souls, valuable to artists and novelists, the myth of the Newfoundland dog as a candidate for the Monthyon,[5] saving his master's child who fell into the lake while picking a bouquet of forget-me-nots, but conscience dictates that I state the truth.

We ask the minister if it is possible to obtain any fur, and he offers to take us to one of his parishioners, who may have some. We go out together and, as we leave, I glance through the windows of the schoolroom, where about twenty girls are seated on benches, studying. In northern countries, like Iceland, Sweden, and North America, children's education is much more complete than in the warmer countries; winter creates leisure time and prevents them from wandering; this meteorological reason is probably not the only explanation.

As soon as any traveller arrives in Newfoundland, he quickly sets out in search of fur, but rarely succeeds in finding any. Animals are abundant on the island: muskrat, arctic hare – whose fur is difficult to distinguish from that of the common white rabbit – red fox, beaver, otter, and finally silver fox, which is very expensive. I am not talking about seal, which is hunted in the winter; the seal hunt is the main winter occupation and Newfoundland's only industry. The animals are shot or trapped by fishermen all along the French Shore, and the pelts are nailed to a board to dry and cure until the beginning of spring. As soon as the ice has melted, fur traders sail along the coast and trade seal pelts for food and household utensils. The pelts are then collected by one of the St John's merchants and sent either to Canada or to Boston. As a result, it is very difficult for a traveller to obtain furs, with the exception of hare and fox, which are common and practically worthless.

Led by the pastor, we arrive at the merchant's, who sells groceries, hardware, and novelties and is also postmaster. At this time, all he has is a bad otter pelt, which we quickly decide not to purchase; but we do have a thorough look at his store, which reminds me of those in the frontier towns in the United States like the town of Crow Wing, Minnesota, in the upper Mississippi in the olden days, an accumulation of all sorts of objects: clothing, thick woollen fabric, axes, barrels of flour and pork; from the ceiling hang gloves for fishing cod, boots, oilskins, and rain hoods for sailors; boxes of nails, crates of tea and brown sugar, saws, knives, glass, pottery, and dishes. You are taken aback by the acrid smell, which in spite of the absolute cleanliness of the store, is given off by all of these various and sundry articles. This smell is particular to stores in North America. Each country has its own; there is a Spanish smell, an Italian smell, an Algerian smell, a particular odour that impregnates the air of a given country and penetrates the inhabitants' attire, in the same way that the language that is spoken is perceived by the ear, and the eye takes in architecture of buildings and monuments. The merchant is not at all busy since there is no schooner preparing for departure and the fishing is poor, the cod having left the shores over the last few years to congregate on the banks. The result is that elsewhere on the island there are no fish, but there are too many on the south coast. Such is the way of the world; people are never content, even the most fortunate who, for want of another source of misery, complain about having too much.

We return by the same path. When we step on the boat that is to take us back on board, a woman asks us whether the doctor is with us. Fortunately, he is, and since our excellent doctor is always ready to give generously of his talents and devotion, he immediately goes to the sick person's house. These people are familiar with the *Clorinde* and impatiently await its arrival each year because they know that the ship's doctor cures sickness and requires no payment. The illnesses result from thin blood, due to the poor diet composed solely of fish, and to the humidity and long confinement of winter. The frigate does not stay long enough in one place to treat or operate on all those who need it, so the doctor can only dispense advice concerning health and hygiene. People never fail to ask for brandy, their universal remedy, and because nearly all the inhabitants belong to temperance societies, nowhere on the coast is it possible to obtain a bottle of wine or spirits. All told, they are no worse off for it, since drunkenness wreaks the most terrible of ravages wherever it strikes, particularly in northern countries, where such a scourge originates from the need to warm oneself and the ease with which alcohol can be absorbed as an excellent respiratory beverage and tonic. Religion, for all its direct and indirect consequences, is the only force strong enough to fight drunkenness. The English and the Americans, having learned by sad experience, are firmly convinced of the indefatigable perseverance of the clergy who have devoted themselves to this humanitarian task. One cannot but praise their courage. I am aware that in France, the opposite opinion prevails; the idea of absolute rest on Sunday, the compulsory closing of cafés and inns, and the related fear of causing a scandal, inspires protest and criticism from my fellow citizens.

Everything fits together; the old proverb "Whoever wants the end, accepts the means"[6] is profoundly true. It is not a question of theory or of a facile commonplace concerning liberty; it is a practical question. I have visited many countries where drunkenness is a problem. I have consulted with the most serious, the most experienced, the most learned people there; and I declare, along with them, that I know of absolutely no other means of fighting alcoholism, that vice which, when it afflicts a nation, corrupts both body and soul.

Ingornachoix Bay, Port Saunders, and St Margaret Bay

From Bonne Bay to Ingornachoix Bay is not far; we sailed at four-thirty in the morning and dropped anchor at Port Saunders at four o'clock in the afternoon.

Our ship hugs the coastline, which at first continues to be equally as majestic, its tall mountains dropping straight into the sea. In places, the wall is broken by a fissure at the bottom of which flows a stream. The Martin fissure is the most remarkable of all; it is an opening that is about twice as high as it is wide at its widest point and whose base is so narrow that the sides seem to meet and converge to a point and the walls to lean together. In the space between, you can see the mountains, which continue to rise up in a series of increasingly higher peaks. In the foreground, there is a low beach, formed by the accumulation of debris carried by the current and spread out by the action of the waves, presenting what geologists call alluvial debris. The place still possesses certain typical traits found in fjords.

Beyond the mouth of Portland Creek, the mountains stretch farther; as we go north along the shore, they rise up inland but soon disappear in the distance. The low-lying section of the coast turns into a series of rounded hills covered with fir trees, and the scenery becomes rather monotonous. We finally enter Ingornachoix Bay and, near Port au Choix, behind the trees, to the right of Point Riche lighthouse, we see the masts of the *Drac* and the *Perle,* the two other ships which, along with the *Clorinde,* make up this year's French Naval Division in Newfoundland. We pass Keppel Island and drop anchor in the bay at Port Saunders.

Newfoundland (West Coast). Eroded limestone,
Ingornachoix

Newfoundland (West Coast). Lobster plant, Port Saunders

This is lobster country; there are an absolutely prodigious number of these animals living in the cold water here, and in spite of being fished, their number does not seem to decrease. All the canned lobster consumed in the world is fished on this coast. In Port Saunders, the fishermen claim to take between 5,000 and 6,000 per day, and since the workers say they prepare 3,000 tins per day, each containing the meat of two or three lobsters, the two figures concur. In Bonne Bay, one lobster plant cans 12,000 lobsters each day in July; but in August, when the shells have less meat, they limit themselves to 6,000 and in September to 8,000. A half an hour after our arrival, the dinghy leaves and returns immediately with three huge buckets filled with beautiful lobsters, snapping their claws and moving their legs, climbing up, crawling about, and making a hissing sound that is particularly pleasant to the ears of the crew, who look forward to a veritable feast this evening.

The canning industry in Newfoundland is certainly destined to undergo considerable development, especially if certain intelligent and strictly enforced regulations are able to prevent the wastage that is all too easy, given the present abundance. First the animal must be fished. A man goes out in a boat with a bucket, a tub filled with bait, which most often consists of the entrails of freshly cleaned cod. He heads out about a hundred metres from shore toward his lobster traps, which are set out along the bottom attached to a rope with a floater at each end. The trap or pot is a half-cylinder one-and-a-half metres in length, three semicircular staves connected by a net to which is attached a wooden ring large enough to let a lobster through and which is drawn toward the inside of the trap. The bait is placed inside, on a stick hanging from a string, or inside a small, open box. In the latter, the bait keeps longer, since the unfortunate lobster cannot eat it. In exchange for his life, he does not even get a good meal. The fisherman hauls each trap into his boat, lifts one of the ends of the net holding the lobster while closing the other end, or opens a hatch in the side, and takes the lobsters out. Then he tosses them into the boat, puts more bait in the trap if there is none left or if it is bad condition, and puts it back in the water. He empties all his traps one after the other and brings his loaded boat back to shore, and his catch is piled on the wharf, where his victims are counted.

The wharf is built of fir trees and stretches out from the shore to where the water is deep enough that schooners can dock and

Newfoundland (West Coast). Workers at lobster plant, Port Saunders

load the canned lobster, which will be transported to Halifax, and unload the supplies needed by the twenty-eight men and women who work in the plant, which is connected to the wharf and is made of two large one-room or two-room wooden houses or stores.

In the first room, the lobsters are cooked in two large vats heated by a wood fire, fuelled by logs cut in the forest, no more than a hundred metres away. Each is filled with seawater and contains about a hundred lobsters, which are cooked without any seasoning and collected by a worker using a net placed on the end of a long handle and resembling the landing net used by trout fishermen. The floor is covered with a reddish-yellow purée, the sauce that is so sought after by gourmets. Fortunately, it is not collected but is simply washed down with water and ends up running into the ocean near the lobster plant, where the bottom is teeming with small grey plaice attracted by such an abundance of food. The lobster are dumped on long boards where they drain and cool, then placed in even rows with their tails all pointed in the same direction on four benches laid out in a square, with an open space in the middle. When I visited, there were three or four thousand lobsters laid out in this way. Such a colourful sight would undoubtedly inspire an impressionist or a *luministe* painter,[1] except that such a rubicund congregation, such a conclave of cardinals of the sea, of course all deceased, would necessitate a substantial expenditure on vermillion or Saturn red. Each lobster is taken and laid on a table; with a cleaver, a worker crushes the claws and the shell, removes the meat and immediately brings it to the women's workshop. The shells are thrown into the sea, where they pile up and form a red fringe around the wharf; the entire coast is lined with them, and they give off such a nauseating odour that it would not take long before you would be deterred from canned lobster forever.

The workers place the pieces of meat in the cans and weigh them; another woman packs them down using a disk on the end of a metal handle. The can is then passed along to a worker who very quickly and skilfully attaches and closes the cover by a series of small blows with a hammer; he then passes it to the welder, who seals the cover, leaving only a small hole in the centre. The tins are then carried to the third room and placed on large metal plates with holes and placed in vats and boiled in fresh water. The lobster is recooked, the air inside the tin dilates and escapes, and then, with a drop of solder, the hole is sealed. The seal is airtight, and all that remains to be done is to coat the tin with varnish, wrap it, and pack it in crates for shipping.

All these operations are carried out quickly. The tins are manufactured locally, punched from sheets of tin; the seawater for cooking is pumped up from the end of the wharf; the fresh water comes from a spring two hundred metres away in the forest, collected in a barrel and brought to the plant in a simple gas pipe. The men and women workers are housed in three wooden houses; during the week, they work until sunset and then rest. On Sundays, they walk along the beach toward Port au Choix. Two by two, each with his or her "better half," as they say, they sit on the grass near the harbour's entrance and talk while looking at the ocean. I do not know what they talk about, and I would not dare guarantee that the day is spent in contemplation only, but from a distance, it is a like an eclogue or pastoral poem. Two or three people live at the fish plant all year round to guard the equipment. The season lasts five months, which means hardly more than four months of actual production, since workers need about one month to travel to and from Bonne Bay and the fishing villages of the south coast of Newfoundland, and even from

Prince Edward Island. The simplicity with which the plant is constructed is such that the entire establishment could be dismantled and taken away in three days. The reason for this is primarily economic: since it works fine as it is, it would be useless to build anything luxurious. Also, it is more in keeping with the good old Treaty of Utrecht, which prohibits both the English and the French from erecting any permanent buildings on the French Shore.

A walk from the lobster plant to Point Riche, following the north shore of Ingornachoix Bay, is interesting for a geologist wishing to observe the endless work by which the earth is transformed. There is a cliff at the entrance to Port Saunders; the top is covered with fir trees and overlooks the beach, awaiting the moment when the frost has completed its destruction. The frost digs relentlessly away at the foot, which the waves continue to beat. One day the frost will deprive the cliff of its support and it will fall under its own weight. Twice each day at the beginning and end of winter, before the entire coast is trapped in the ice, the tide covers the rock, saturates it, and recedes again and abandons it to the action of the cold air, which causes it to shatter. The tide returns and carries away the debris that has accumulated; in this way, the hills crumble and diminish slowly and are replaced by a wide beach covered with rough-edged rocks. Occasionally, a harder rock is chiselled, rounded, pierced, shaped in the strangest fashion, in the form of a needle or a mushroom. This rock, which resists erosion, bears witness to a form of energy of which we in our temperate climates have no idea. Farther on, an enormous limestone deposit stands in the ocean, carved out by the waves. Over a distance of several kilometres, this mass has become like a huge wave itself, with parallel rows and ridges carved into it by the sea. The spray keeps replenishing the pools of seawater, and lines of large red blocks have been deposited after having broken loose in Labrador and been transported here by the ice. The rock breaks apart easily and reveals many fossils, creatures that once lived in the sea and whose bodies have been fossilized for millions of years, only to be soon dissolved and returned once more to the ocean, to become part of other animals in the future. Each indentation in the shore, especially the north-facing side, is covered with these erratic blocks; in front of them there is a band of trees, their roots twisted and intertwined. These trees have been deprived of their bark, giving

Newfoundland (West Coast). Eroded cliff, near Port Saunders
(northeast of Ingornachoix)

their trunks a polished look in the silvery sunlight. In nature, movement is never-ending and matter is continuously transformed and retransformed, forever changing and contributing to an eternal cycle. Time does not exist; while we are busy counting the years, the weeks, the days, and even the minutes and the seconds that pass, these immortal things in nature peacefully contemplate the centuries as they go by.

Near the end of our stay in Port Saunders, the fog returns and rain begins to fall, the heavy grey sea as smooth as a mirror. The raindrops make large bubbles, which float along in the current for a few seconds before suddenly bursting. Along the water's edge, the houses near the lobster plant stand out against the endless expanse of forest. The concave cliff of Keppel Island is visible, and with the aid of a telescope, it is possible to distinguish a seal frolicking near two seagulls, which watch him carefully while he pushes his thick body and swings his round head. What do animals think about? When their hunger is satisfied and they do not fear for their lives, must they not use their leisure time to recall memories and organize them, to draw conclusions that will later amount to experience? Creatures belonging to the same species speak the same language, while those of different species have more limited conversations, though they nonetheless understand one another. I observe the seal and the two birds, and at the same time on the bridge a worker from the lobster plant who, without knowing a word of French, converses with our sailors, each of them using gestures and understanding the other. Animals communicate by making sounds or, like ants, by producing simple vibrations that are inaudible to humans. Every sensation is a vibration, which, depending on its volume and on the nature of the organ that perceives it, is either a sound, a glow, an odour, or a taste exactly like the colours of the spectrum for us humans. Each vibration would be coloured, warm, or actinic, or both coloured and warm or coloured and actinic, according to its wavelength. Certain animals hear what we see and see what we hear and smell. Using the wavelength, science will find the common measure between all forms of sensation, no matter which sense organ they are perceived by. We will then realize which analogies we sense but are unable to define; we will grasp strange associations detected by certain individuals but denied by others who do not perceive them and that appear real – often all too

Newfoundland (West Coast). Eroded rock, Ingornachoix Bay

real – to those who do, like vague harmonies resembling long forgotten memories that blend impressions of entirely different sorts: sounds, colours, thoughts, the look of the *Venus de Milo* and lines from a play by Racine, the rhythm of a Chopin waltz and a painting called the *Larmoyeur*[2] by Ary Scheffer,[3] a Beethoven sonata and the fragrance of a certain flower. Let us follow the advice of Jean-Jacques's Venetian[4] and not study too much mathematics, or at least not only mathematics.

An unfortunate fisherman has come to have his arm amputated by our doctor. Hand lines are often used to fish cod; the line is wet, and if the fisherman has a cut, the salt water will cause it to infect and to develop whitlow. The men live in dampness and filth, without a doctor's care. Whitlow develops into a wound that worsens continuously until amputation becomes the only possible remedy and the person is lucky if there is a doctor to perform it. The man walked all night to get here; when he arrived, the doctor put him asleep and operated on him. He stayed on board for two days and was fed and cared for; he recovered slightly and left to return to his own boat. A fisherman is paid monthly and receives a bonus of five francs per thousand cod that he fishes or one half a centime per fish. He must continue to toil, otherwise his wife and children will go hungry next winter and his misery will start over again. What a poor unfortunate fisherman! May God prevent his sickness from worsening!

Leaving Port Saunders, we head north and sail between the low-lying and wooded St John Islands. We follow the featureless coast very closely and eight hours later enter St Margaret Bay, a large gulf that is wider than it is long, open to the west and bordered to the south by New Férolle and to the north by Old Férolle peninsula. In the southernmost part of the bay, several islands barely rise above sea level, Île Verte, Île Boisé, Île aux Brousailles, Île Rase, Île aux Oiseaux.[5] All these rugged islands and peninsulas provide evidence of an extensive sinking of this part of Newfoundland. The beach is narrow and adjacent to the thick forest of fir trees. The St Margaret Mountains rise up in the southeast, and the nearest foothills form a large, flat plateau about ten miles away, like gigantic fortifications with bastions. They are composed of parallel, slightly folded layers of limestone distinguishable by their different shades, and they are separated by clumps of trees so that a side view represents a series of steep steps set farther and farther back. Halfway up, an immense scree with even slopes like a pedestal supports the entire base of the mountain. The upper part is bare, but as you descend, vegetation suddenly appears like a line so sharp that it looks to have been painted by a brush separating the rocky and sterile soil from the vast forest. A chain of rounded hills amid numerous glacial lakes reaches back between the sea and mountains. The ground is carpeted with moss and ferns; elsewhere, peat, with trees whose thick branches grow from the ground up and present major obstacles to walking, except along streams. In places, the beach is as much as one hundred metres wide, with outcrops of limestone that have been polished, rounded, or striated; most of it is strewn with piles of rocks of various sizes. Here also the erosive action of the frost is visible.

Some of the grassy islands have wooded interiors and are seldom likely to be under water in spite of being so low in elevation. Flocks of grey-backed gulls with black heads and coral-red legs and beaks nest there. As soon as they notice our boat approaching, they take off and fly around us, making their raucous cries. Nests are scarce, but we find a few where the hen has laid greenish eggs with black spots in the moss on a bed of feathers picked from her own coat. Nothing is spared from the hunter's gun; the unfortunate creatures are greeted by a volley of blasts that leave quite a number flopping and gasping around us. It is a pity to shoot birds that are so tame, especially since they are not even eaten. The sailors tell us that their meat is very tough and disgusting in taste, and we cannot doubt their word, since the monotonous diet on board does not make it difficult to appreciate any new dish. The gulls roost here on the rocks. In the morning they fly away and hunt for food, gliding over the sea, where they occasionally dive to catch a fish and carry it off in their beak; in the evening they sleep huddled together. In summer, they are not bothered by the huge steamers that sail along the Labrador coast, nor by the sailors who are busy fishing for cod or lobster. When winter comes, they are entirely alone with the white mountains, the frozen streams, the silvery fir trees weighed down under the snow, the bay which resembles a level field. Far in the distance, the furious waves break against the ice in an unrelenting attempt to crush it, and the ice, frozen solid again by the cold, piles into enormous blocks, only to be broken again. The gulls are the only living creatures in an otherwise dead nature; in the fog, their cries blend with the wind and storms, and when the weather is calm, the moon's rays shine and the stars twinkle over the vast icy plain.

In this region of the country, winter must certainly be marvellously beautiful; but this evening, leaning over the rail, I admire the huge, red disk of the setting sun. The sea is a very bright shade of bluish-green; small waves lap against the side of the frigate. The light shines on the delicate outline of the fir trees which stand out against the sky, where scattered purplish clouds slowly darken the upper portion while, near the horizon, oblique pink and green bands shine with the brilliance of stained glass in a church. The masts of a schooner at anchor, whose hull is lost in shadow, stand as straight and taut as wires. The sun goes lower, the sea becomes steel blue, the sky takes on the same shade as the sea, then becomes purple; the metallic colour of the water fades and the blue deepens as darkness begins to fall, first leaving a thinning ray of gold, now only a burning trace soon to be extinguished while the cool evening air spreads itself around; the winds dies; night has fallen, and the earth, the sea, and the sky now sleep in silence.

7

St Lunaire, Croque Harbour,
and the Mosquitoes

We encounter our first iceberg on leaving St Margaret Bay. It is about twelve metres long and thirty metres wide, with a valley in between the two mounds, one pointed and the other shorter. No matter what side you examine it from, its outline is nearly always curved. At the highest point, it is a dull white, but nearer the bottom, it becomes transparent and takes on a glaucous green that a watercolour artist would try to render using light Prussian blue with a touch of ultramarine. The ice has a series of bluish or slightly grey layers, parallel to or more or less inclined toward the water. There are hollows or caves where the light draws the most incomparably soft shadows. The overall effect is charming, astonishing, and terrifying at the same time. We fire a shell at it; after the bang, we see the shell penetrate it, causing a tiny avalanche to slide into the sea. The seabirds perched aloft fly off in a flurry like tiny black dots, and the iceberg quietly continues on its way southward.

As we sail farther along the Strait of Belle Isle, the number of icebergs increases. We already count eighteen all around, though the winter has been remarkably mild this year. They break away from the great glaciers of Greenland and drift down the Baffin Sea, following the Arctic Current, which takes them along the coast of Labrador and the east coast of Newfoundland. Few of them cross the Strait to make it to the vast waters of what is improperly named the Gulf of St Lawrence, located between southern Labrador, Canada, and the west coast of Newfoundland. During their trip, the icebergs melt as they meet warmer air and water, and when by erosion their centre of gravity alters, they roll over. One of their extremities, having become lighter, emerges while the other sinks below the surface. Since the waves wash against the lower portion, they carve out a rounded mould all around, leaving a groove that is characteristic of "ice mountains" in these waters. Their form depends on the latitude; in the north, near their point of origin, they are angular and sharp edged. The longer the trip, the smaller they become; some even disintegrate completely and break into a thousand pieces, covering a large area of the surface of the water with their floating remnants. On reaching the Newfoundland Banks, they meet the warm current of the Gulf Stream and are annihilated, while cooling the moist air around them. Since they are lighter than water, in order to float, icebergs must keep most of their volume submerged, approximately six-sevenths. Sometimes this hidden foundation scrapes the ocean floor; usually it follows the direction of the lower currents, and if the above-water portion is pushed by favourable winds, the berg crosses the Gulf Stream. It is said that some have been sighted as far as the Azores. We spot them on the horizon but do not fully realize how gigantic they are until the schooner passes near them, like a tiny black dot beside an enormous white mass. Most icebergs measure fifty-five metres above sea level, which makes their total height four hundred metres. Their grandeur, their strange yet beautiful shapes, their soft shades of colour, their graceful movements, strike the spectator with awe. At dusk or at dawn, in the faint twilight as they drift idly by, it is understandable how simple Scandinavian sailors imagined them with transparent caves inhabited by mermaids, blue-eyed girls of the northern waters, plaiting their long wet hair and accompanying their prophetic singing on silver harps, protected by terrible sea monsters and the awful *kraken*[1] or giant squid, with its threatening horns, or by sea serpents rearing their heads and horrible manes with their scaly bodies undulating.

The Labrador coast is too far to be seen from here. As we continue through the strait, the Newfoundland coastline becomes barer and barer. The Cape Norman lighthouse at the northern extremity of the island is built on an arid plateau above a beach strewn with rocks. The *Clorinde* glides past Pistolet Bay, where a magnificent iceberg has gone aground with its half-melted top looking as if it were veiled with lace. We round Cape Onion, and to our left the Strait of Belle Isle stretches away into the light fog.[2]

Newfoundland (Northern Peninsula). Village of Anse-à-Bois
[Wood Cove] Quirpon

We pass Cape Bauld and then Quirpon Island and head south.
The features of the coast are difficult to distinguish from a certain
distance; bays and headlands appear to blend together when
viewed from afar and when the clarity of detail that would normally permit you the determine which is nearer and which is
farther is blurred by the varying visibility. Certain features are
masked by the fog while other distinctive traits appear sharper
and therefore closer. The high rocks against which the waves
break are concave in places as a result of the combined action of
tide and frost.

We anchor in St Lunaire Bay, where there is another example
of land shaped by the cold climate. Its irregular contours, sparse
islands, and numerous coves with headlands receding behind one
another all bear a certain likeness to our Morbihan;[3] but its vegetation, made up of spruce and dwarf juniper, is more reminiscent

of the islands of St-Pierre and Miquelon. In order to walk, you must undertake exhausting gymnastics again and tread upon the thick tops of bushes where your foot rebounds until it eventually sinks into an invisible opening.4 The bay is surrounded by mamelons5 separated by wooded valleys; at the head of the bay, where the ship is anchored near the mouth of a stream, there lies a wood in the middle of a few small, absolutely delightful meadows. At sea level, lamellar schist, some of which contains pyrite crystal, is being constantly eroded and washed away by the tide; higher up, more resistant blue sandstone intersected by veins of carious quartz forms a steep slope plunging down into the water. At the top of the hill, the rock juts up through the moss and, since it is exposed to the elements, crumbles into rough, angular stones. The debris collects in hollows, where water also accumulates along with minute fragments of plants and bushes, which disintegrate and transform the vegetable and mineral matter, producing a spongy mass that becomes water-logged and eventually covered by moss or shrubs. This explains the formation of the heather or peat lands, so common to the island of Newfoundland, where the water flows red with tannin. Everything in nature is inscribed and visible: the present is a direct consequence of past events and the immediate cause of future ones, to the extent that every natural phenomenon is predictable. It is the aim of science, through arduous effort, to seek to formulate the laws that make such predictions and understanding possible. The vegetation binds with the composition of the rock which sustains it; there is no better place than here to study the bonds that exist between the mineralogical and chemical composition of the soil and the varieties of plants nourished by it.

We receive our mail, which the steam launch went to fetch in the very humble but very blessed post office at St Anthony. For a long time now, our letters have been awaiting our arrival, and now at last we are watching for the launch's smoke to appear behind Nymph's Peninsula as she brings them to us. Finally, she approaches and moors alongside, and the cherished mailbags are carried up the ladder and taken to the captain as we make way to let them pass. For several weeks we have been awaiting the mail without thinking about it too much. Now that we need wait only a few minutes longer for news from those dear to us, we become terror-stricken and impatient and accuse the quartermaster of being slow as he opens the door every few moments to pass out a package of letters or newspapers and quickly returns to get those that have just been sorted. We share the treasures, and once we have received them we carry them quickly away to some corner or to our cabin or to the mess to enjoy them in peace. The mere sight of a familiar signature or handwriting is enough to calm the fear in our hearts; we feel reassured even before opening the letter we hold; for a moment we hesitate to choose which one to read first, and the feeling is most enjoyable. The envelopes are torn open, the letters read; everyone is in good health and thinks about the absent person, calculating in advance how long it will be until his return. When this need for affection is appeased, the mind, which has its own appetite, also wants to be satisfied, and so we break the paper bands binding the packages of newspapers together and begin reading everything pell-mell: politics, the caricatures in the illustrated magazines, news of friends and colleagues who have been appointed to new positions. Everyone reads aloud to his neighbours, who are not listening, shows to others, who are not looking, letters that are covered with stamps, postal curiosities that were sent months ago, that have been sent to China, back to Europe, visited Brest, Cherbourg, or Toulon, travelled to Gabon and back to France and finally arrived in Newfoundland via the West Indies, following the sailor whose whim and wandering career fled before the letter in spite of himself. He finally reads old news with a smile on his lips and sometimes a tear in his eye. I see one man hold in his hand a letter from his father, whose last breath he watched him breathe before sailing after a brief stay in France. The letter had been sent more than a year earlier, had gone around the world, and in poignant irony, spoke of joys and hopes that were now buried in a tomb. How cruel the life of a sailor is!

Leaving St Margaret Bay, we pass Amelia Cove, where there are a few shacks used in winter by the inhabitants, who hunt furs and in summer fish cod and salmon. The weather is splendid but cold. On our left in the distance we sight Brehat Shoal, which is eroded by the sea but rises straight up like swords in the foam. As soon as we round Cape Dégrat, we enter Croque Harbor, where we must stop. In fact, we return several times in order to put the cattle we have transported from St-Pierre to pasture. The poor animals have no sea legs and are tied on deck, where they have been

Newfoundland (Northern Peninsula). Head of Croque Harbour,
south of Hare Bay

fed dried hay drenched in salt spray and have suffered cruelly
during their stay on board. Their meat is fibrous, tough, and
tasteless and wreaks a terrible vengeance on our teeth. They are
unloaded quickly using two straps and a block and tackle, put into
the water, where they are quite able to keep themselves afloat,
tied together by the horns, a floating herd of red snouts snorting
in fear, and towed by a launch toward La Forge, where they scur-
ry out of the water. As soon as they feel solid ground under their
hooves, they shake themselves, smell the green grass, lower their
heads, and begin grazing. The joys and sufferings of animals are
so short-lived! La Forge is a cabin where three men live and for a
couple of months play Tityrus,[6] Melibee,[7] or rather Robinson
Crusoe and enjoy the leisure time left after tending the cattle by
growing radishes, cress, and cabbages brought from France, which
they plant in a small, enclosed garden.

The sight of Croque Harbour and its belt of wooded hills is a
welcome relief from the sad, arid look of the Northern Peninsula. The vegetation nonetheless suffers from the proximity of the
icebergs and is not as thick as on the west coast, making it easier
to walk in the forest. It is certainly not with the greatest of ease,
however; you still have to climb over or slide under fallen trees;

Newfoundland (Northern Peninsula). A guardian and his family

occasionally, you reach higher, more open ground, where grass grows among the disintegrated rock, and where you have a view of the entire area. There, you take your bearings and choose the direction you will try to follow once you go back into the trees during the descent. In this fashion, we travel from La Forge to La Genille,[8] which is a group of four or five cabins at the head of a small cove where Patrick,[9] the winter guardian of our station, lives with his family. Near the ocean he has set up a small stage where, for lack of anything better to do, he is busy drying capelin. From La Genille to Cape Dégrat, in spite of its height, the terrain is swampy with large clumps of *sarracenia*[10] growing, their horn-shaped leaves spotted with red. Their strange, greenish flowers appear to be cut out of plate zinc. Facing La Genille at the opposite end of a deeper harbour, which seems to be guarded by an iceberg that is grounded and being demolished by the waves, lies

Newfoundland (Northern Peninsula). Cemetery, Epine Cadoret,
Croque Harbour

Newfoundland (Northern Peninsula). Observatory Point, Epine Cadoret,
Croque Harbour

La Plaine, a fishing establishment occupied by a French crew. The cod fishermen there also have cause to complain: the fishing is poor and their captain just died. Yesterday, he worked all day and in the evening felt out of sorts. His companions heard him get up in the middle of the night and go out; this morning they found his body lying face down. He had formerly served under our own captain;[11] since he was a child, he served first as cabin boy, then sailor, and finally skipper and carried out his duties quietly and simply, without wasting any time and without a word. When his time had come, he laid down his knife and died. He will be buried tomorrow in St Julian's, where his boat is. All that his men have to say about him is that he was a good man, and that funeral oration is as eloquent as any and truer than most.

Croque has its own cemetery with a large white cross at the entrance to Epine Cadoret in front of Pointe de l'Observatoire, where the officers customarily carry out their observations.[12] Sailors belonging to the French station and to the English station who die in this area are buried here. The graves are hidden in the tall grass covered by wildflowers with only the heavy, solid wooden crosses, carved and painted black by navy carpenters, visible in the greenery. Sailors and officers alike, French and English, Catholics and Protestants, a certain Villeret de Joyeuse between a French quartermaster and an English novice, sleep side by side in absolute equality. Why is death more terrifying in the city than in the country? In this solitude, beneath these wild mountains where few fir trees grow, it seems serious, melancholy, but devoid of horror. Is it because whatever is fashioned by the hand of man, the products of our industry and our art, all sing the praises of life and protest against death, whose very name makes us tremble? In the bosom of nature, life and death form such a close alliance that they are inseparable; if life evokes death, death also evokes life; one follows the other, each only an episode; on the ground covering a tomb, you tread on last year's dried leaves, with the green leaves and sap-filled buds bursting with life above you. The moment when a life reaches its end is the moment hope begins anew; winter follows summer but makes way for spring; evening holds no cause for dread because we know that morning will soon come. The farmer, the sailor, and the savage all accept death more easily than the city dweller; they come to the end of their lives while he approaches death.

Of the entire French Shore, Croque Harbour offers the best possibility to make an easy or rather less difficult long walk inland, to a lake at a distance of a half a mile. You go ashore at the wharf at La Forge, follow the beach to the left, and as soon as you reach the head of the bay, you arrive at a stream where water cascades onto large rocks. Before flowing into the sea, the water becomes still and gathers in a natural pool, which serves as a washing pool for the crew. What a pleasure it is to see the sailors wash their clothes on the rocks, bathe and splash around in the fresh water, and enjoy stretching out on the grass. They are scattered around the pool, and their movements bring life to the otherwise sad landscape. Men appear as tiny dots and make the mountain seem even bigger than it is because they serve as a comparison, a scale, as painters say.

I begin to ascend the path along the left bank, enter the woods, and without running the risk of sinking into the bog, am able to advance and admire the green vault of the overlapping fir and birch branches overhead. The stream has no banks but runs through the brush; in places where large boulders have fallen down and been carried by the current, it flows over them and covers them with white foam or abandons its bed, changes direction and follows the path. Several primitive bridges have been put in place to cross it; the first one is called the Marie-Louise Bridge and is made of fir tree trunks simply laid side by side. They were probably in excellent condition two or three years ago; today they are half rotten and shake in a rather worrisome way when a walker steps on them. The moss is light green and dotted with violets, irises, buttercups, and pink heather with flowers resembling tiny bells. I arrive at the lake, where the still waters reflect the sky and the clouds. Everything is calm and tranquil, no sound of birds, not a ripple, not a rustle in the leaves. I am reminded of the region at the headwaters of the Mississippi where I travelled sixteen years ago.[13] It was in terrain like this that I camped every evening after a long day's trek through the swamp, exhausted and worn, my shoulder sore from carrying the tripod of a surveying instrument. We would stop, cut down a few trees and quickly start a fire with birch bark, which ignites even in the damp, the flames from the resinous wood reaching high. We heated and ate our meagre portions of beans and salt pork, carried on our backs over many miles of marsh. The tents were put up over a layer of cypress branches, laid horizontally to prevent us from sinking into the moss, which was as wet as a

Newfoundland (Northern Peninsula). La Genille, Croque Harbour

sponge. We would lay at one end a small pile of clothes and some leaves as a pillow, roll ourselves in a blanket fully dressed, and fall asleep. Sometimes a storm would pass over our camp, the flashes of lightning so bright they dazzled us even with our eyes closed, the roar of the thunder joining the cracking of branches and trees breaking and falling on one another, producing a shock that we could feel in the ground. But whether the night was good or bad, during our sleep, the mattress made of branches slowly gave way, and we sank gradually down into the water without realizing it, first our feet because of the clothing, then our legs, followed by our torsos, and still we would sleep because our fatigue was so great; finally our neck would be submerged and we would sleep through it all.

Fortunately, this would not happen until morning. The swamps and marshes of the Mississippi are such that, in spite of all our precautions, they require six or eight hours to swallow up the body of a man lying down. This was our only source of entertainment. The first man to wake up watched his companions, and if the poor devil slept for a few moments more, the water reached his mouth, and suddenly the snoring became a long gurgle, followed by a hiccough and a fit of coughing that seemed to us funny enough, in spite of our misery, to make us laugh. It happened to each of us in turn. Of course, it would not last long; we promptly got up and broke camp after a breakfast of beans and salt pork, and we set out again through the brush, dreaming of the ideal life of those who could sleep as much as they wished in a dry bed at least once a week. For me, in the wilderness, there were moments of enjoyment when I would nestle on the shore of some unknown lake and at nightfall see deer, stags, and elk with their huge flat antlers[14] come out of the forest, cross the shadows of the tall fir trees, come to the water's edge, and drink, lifting their heads and pricking their ears at the slightest noise. A lot of fatigue, little sleep, and not much sustenance – life was difficult, with its rare joys and such frequent miseries, but it left me with memories whose pleasantness is beyond words. Some years later in Paris, my work required me to walk through the Jardin des plantes every morning, at a time when no visitors were present; the walkways were so deserted it took very little imagination to feel like I was absolutely alone and in my own garden. One day, I recognized a Norwegian pine, with its wide branches looking firm and majestic. Of course, this poor domesticated specimen was not equal to its kin on the Mississippi, and it resembled them only as a penny does a gold piece, but I felt as though I were looking at an old friend and confidante with whom I had shared many troubles, and I was struck with deep emotion. Today, it is no longer a mere trace of my past life that I re-encounter here but almost my life itself; the scenery is identical, I am happy, but too much time has passed, and I am not as moved as I would have thought I would be or as I would have liked to be. Alas, it is the spectator who is at fault, not the spectacle. Youth has flown, the fairy who changes everything into diamonds with a touch of her wand and who still smiles after even the most bitter of tears. Of what importance is the past, since it no longer exists? A young man leaps forward thinking only of what the future holds for him, and because it belongs to him, the weather is always fine. The years follow one another; more than one of them has been a burden. Each of us counts the weaknesses in his body and the scars in his heart. Yesterday served to teach us where we may bleed red blood, and so we fear tomorrow. No, we must not look back too often to gaze upon the faded flowers that we saw alive and fresh. Heads and hearts up!

And so I wonder whether it is even possible to let my enthusiasm run wild while throughout the entire walk I am devoured by mosquitoes, which swarm around my face. In spite of my calico hood and gauze veil, with the lower portion covered by my buttoned-up clothes, in spite of my woollen cap pulled down to my ears, in spite of my determination to keep my hands in my pockets and not remove them for any reason, they penetrate openings known only to themselves and provide me with only too much proof of their presence. If I were a poet and heroic poetry today were held in as much honour as it once was, when speaking of mosquitoes, I would certainly begin by invoking the gods of desolation, anger, rage, fury, and despair; I would try to avoid overlooking a single one of them and would beg them all to come to my assistance and would depict, with a seriousness at least worthy of the subject, the suffering and torture these miserable little insects inflict upon the unfortunate summer traveller in the northern regions of North America. It has been claimed that the United Sates and Canada are not alone in possessing such a privilege and that the same scourge exists in all northern countries, such as Lapland, Russia, and Siberia. Zoologists, if they wish, may take care of elucidating this geographic question of natural history. The miseries of others have in fact never seemed to diminish my own suffering: I survived the mosquitoes of the United States and Newfoundland, and now that I have done with them, thank God, I feel the same pride as an old soldier I once knew who had seen the Russian retreat and who lived to tell about it.

You hear people talk about the mosquitoes in hot climates; those are inferior versions of mosquitoes. In Algeria, Spain, Italy, the south of France, you meet only amateur mosquitoes, innocent mosquitoes, or mosquitoes with delicate constitutions, artistic and musical mosquitoes, at times boring, especially for people who like more varied melodies. In those countries where orange trees blossom, when you fall asleep at night and one of those mosquitoes happens to enter your bedroom, it is satisfied to announce its presence by a simple murmuring; it hardly does more than take a few bites that leave only a slight swelling in the morning. I have heard people who have travelled in the south and who claim to know mosquitoes. They are mistaken! You must come here from July until late September to be able to assess the amount of open or repressed anger for which one of these minute creatures can be responsible. Of course, the word "one" is an expression, a rhetorical flower, and you could say one thousand, one million, one billion, one billion billion mosquitoes, *blackflies, gnats, hornets, wasps,* or any other bloodthirsty mob of ferocious brutes, winged tigers that gather in battle formation, rally and attack as soon as you touch land or enter the woods or whenever you stand still to draw or make notes, buzzing around your neck, behind your ears, on your temples and on your hands, totally indifferent to the fate of the comrades who have been crushed and flattened by the hundreds. In warfare, every man wounded or killed is an enemy soldier out of action and therefore one less to fight; for mosquitoes, it makes no difference; a vigorous and well-executed slap on your own face may well crush a certain number of them, but *uno avulso non deficit alter*,[15] his colleagues close the ranks, sound the charge, and launch another assault.

For specialists, there are mosquitoes and mosquitoes; for someone walking in the woods, all mosquitoes are alike. There is the true mosquito, with its long, graceful body and slender legs; its head is small and black with a hairy corset, and its mouth is armed with a sharp proboscis, which it is able to stick into the flesh, and which, after serving as a drill to pierce the epidermis, doubles as a tube by which the patient's blood is pumped out; there is also the *simulie*, a small, fat fly about one-eighth the size of one of our good old French flies, grey in colour, stout, with black legs with a pure white stripe around them. It cuts the skin with its mandibles and leaves a wound that causes an unbearable itch that lasts for two days and produces a painful blister, a real injury that requires seven or eight days to heal.

You defend yourself as well as you can, but any method you use will be ineffective. Some try coating themselves with oil, Vaseline, carbonated glycerine, or glycerine with laudanum; the remedy works for a few minutes but then leaves a greasy residue on the face that is as bothersome as it is useless; others wear veils, or chemists' metal masks, others cover themselves with a hood. At first, the mosquitoes seem a little disconcerted but persevere. They search for a hole, slowly and patiently, and if there is one thread of gauze missing, they find the break in the armour and slide in one at a time and carry out their plundering all the better since, once they have had their fill and try to escape in order to digest in peace, they are invariably unable to find the point of exit and so are forced to go back for more. Moreover, the veil is cumbersome. When the weather is hot, you suffocate; when you walk in the woods and try to squeeze between tree trunks and avoid holes, it tears and offers no more protection or clouds your vision so that you lose your footing and step into the hole you were trying to avoid and fall. At the outset, you arm yourself with a good measure of courage and philosophy; you crush one invader but when the thousandth arrives and you get bitten, you become indignant, furious, you beat yourself, swear, wishing that the mosquitoes had only one head, even if it were as big as the towers of Notre-Dame de Paris, so that you could cut it off with a single blow. Everything you try is futile. You may fight a lion or a snake, a panther or a rhinoceros, but you cannot fight mosquitoes; you fall victim to despair until the moment you are safely aboard the ship, which only happens if it is anchored far from land and upwind; you remove your veil and observe your swollen face, burning ears, and temples redder than a lobster out of the pot. You get off with a day of fever and a week of scratching yourself. You hear stories of people who were lost in the bush and attacked by mosquitoes and were driven insane with the pain and unable to continue on their way, fall down and die, victims of these executioners. In the Croque cemetery, there is the grave of a poor English midshipman who died in this way.

Those of us who like to go trout fishing in the streams suffer particularly because of mosquitoes. You may set out dressed in the most comical way imaginable, but it is useless to disguise yourself even as an Eskimo or don oilskins and sou'wester, boots and gloves; you will come back in bad shape. I have seen coves with streams teeming with mosquito larvae. If one insoluble problem ever came to my mind, it is to know what these animals live on if their entire existence goes by without their encountering a creature to bite. After all, hikers are infrequent in northern Newfoundland. Undoubtedly, mosquitoes die of inactivity and bequeath their hunger to their countless brother "mosquit-eers," who will have the good fortune of sinking their proboscis into some unfortunate fisherman, some miserable geologist, or some unlucky sketcher.

The mosquito and the blackfly appear when the weather gets warm and the cold is passed. When will the frost come to kill them off? In the meantime, we are being devoured.

Jacques Cartier Bay and Sacred Bay

*T*he north coast of Newfoundland stretches in the general direction of east to west from Cape Bauld to Cape Norman and has three indentations. The first is Jacques Cartier Bay and Mauve Bay,[1] the second Sacred Bay, the third Ha-Ha Bay. The *Clorinde* will visit this part of the coast.

Jacques Cartier Bay [Kirpon Harbour] extends to Kirpon to the east and is divided into uneven portions by Jacques Cartier Island. This island is barely one kilometre long and does not have a single tree, and the grass scarcely hides its nudity; it is a meagre tribute to the sailor who, after leaving France, was the first to navigate the great St Lawrence River and the entire northern portion of the continent of North America. Fragments of the ship that carried him over the ocean waves, and through the even more dangerous fog, are preserved in the St-Malo Museum. The ship was certainly frail; today, we would shudder at the thought of crossing the Mediterranean in it, but the soul of its captain was great and strong. Some men are driven by ambition, by that admirable desire to be someone, to accomplish something and place themselves in the way of all sorts of danger. They employ all their strength, their hearts, and their minds. A thousand will fall along this road strewn with obstacles and disappear without a trace; rare are the fortunate ones who finally reach the goal they so ardently strive for. One gave Leon and Spain a New World but had bitterness and ingratitude heaped upon him. The other discovered a province vaster than a kingdom and had an island named after him, so small that the most modest of our

Newfoundland (Northern Peninsula). Jacques Cartier Island

farmers would not and could not sustain himself there. And yet, believing that they are entitled to a reward, they all hope to be paid for serving the general good. Their sense of justice is outraged at the idea that they will not obtain that often counterfeit currency we know as glory. Cavalier de la Salle was betrayed by those who were to protect him, abandoned by all, and murdered on the banks of the Mississippi; Jacques Cartier left his name to a tiny mound of earth. What a sad livelihood to be a benefactor of humanity!

Jacques Cartier Island is nonetheless picturesque to visit; there were once several fishing establishments, of which only one is presently occupied. The abandoned admiral's stage is nothing more than a collapsed structure of fir tree trunks resembling the washed-up carcass of a large whale. The Coupe-Souliers ['shoe cutter's'] stage[2] is occupied by the crew of a French brig. Between the two, near a pool of water, are the dwellings of the fishermen and the winter custodians, along with beached boats; houses and boats alike are solidly braced against the strong winds by heavy timbers and are not very different from one another; the same people live in both. The boats are houses that move, and the houses are boats that do not move. When a rowboat is too old to sail, it is turned over, and placed on four short walls whose cracks have been filled with moss; once transformed into a roof, it can still provide years of good service. On the opposite shore, the houses of L'Anse-à-bois[3] are built of boards and not as picturesque, but more even and certainly more comfortable, which is appropriate for the sedentary population living there.

At this moment, the Coupe-Souliers stage is vacant. To enter it, all you need do is lift one corner of the canvas that covers it. In the centre is an enormous pile of salt; on one side, "green"[4] or lightly salted cod are stacked in rows; at one end is the stall with two chairs for the fish splitters and a hole in the floor, which is supported by pilings like some ancient water dwelling. You can see down through to the sea bottom, which is covered with a layer of cods' heads and teeming with hungry plaice and other small fish. In front of the door is the box used to prepare cod liver oil. Now that it is not in use, it is possible to approach and examine it without the risk of suffocation from the nauseous odour it normally exudes. The large box is located at the top of an inclined ramp connecting it to a vat, which is covered by a large piece of woven canvas. The loaded wheelbarrows are pushed to the top of the ramp and dumped into the box. The awful mixture of cod livers, bits of entrails, and debris of all sorts begins to decompose, and foul-smelling liquid oozes through the canvas and falls into the lower vat. Once allowed to sit, it separates into two layers, with the oil floating to the top of the bloody magma; when it has accumulated to a certain depth, a plughole is opened and the oil is drained off into barrels. The few remaining men from the brig are busy drying capelin on racks while cabin boys sitting on the beach rocks remove the heads. Though this small fish the size of a sardine is sold very cheap, it is so abundant that the quantity caught will still bring a little money. The other crew members have left to fish in their boats. They are gone several days and sleep on the boat or in the shelter of a large rock, wet from the seawater or the rain, forever tired. A rowboat goes back and forth between them and the stage, bringing supplies and carrying back the codfish. If the fish are plentiful, all will be well; all the effort, the exhaustion, and the work will be forgotten.

For the last few years, the amount of fish caught has not been adequate. The cod have moved out to sea, away from the land to gather on the banks, and moreover, it is said that the English are destroying the coast with their traps – long, narrow mesh nets which let nothing escape.[5] They set sail from St John's as soon as the sea is free of ice. They arrive at the fishing grounds quickly and set their traps wherever they have a chance of catching anything. They will take any fish, big or small, because they can sell it, while the French fishermen can take only the large ones. It has been claimed that these thieves use the Treaty of Utrecht for their own profit and cause as much harm to the peaceful English inhabitants of the coast as they do to our fishermen. Moreover, for most of them, the winter seal fishery has enabled them the make enough money for the year, and the cod fishery means extra earnings. Instead of fishing methodically, they are not concerned about destroying everything, and as for the Newfoundland House of Assembly, it is awaiting the day when the last Frenchman leaves and it can become the sole master of the island and the waters around it. Only then will it bother to regulate the fishery, for there will always be a few codfish, and each one produces millions of eggs. They will allow the waters to fill themselves up again or they will restock them artificially, if need be, as in Norway or the United States. Frenchmen will have forgotten about the island, and perhaps new treaties will come along. Whatever happens, the cod will become English; the English want nothing more.

What selfishness, what faults, what admirable virtues can be observed in the English; it is enough to look at the most remote part of the globe, the most humble of villages, the smallest group of huts, in order to understand how the Anglo-Saxon race, by its virtues as well as by its vices, whether American or English, has been able to dominate the entire world.

Having landed in L'Anse-à-bois, I then went to Mauve Bay[6] along a path which crosses the marsh, climbs the crest of the hill, and goes down the other side. The walk takes half an hour; occasionally, a greyish-black robin with a red breast flies up in front of me, resembling the French *merle*, although its song is not as shrill, as prolonged, nor as lively. Our *merle*'s song is one of youth and cheerfulness. When he flies into the woods in the morning, among the dew-soaked leaves, he calls out to tell us how good it is to be alive and how sad we must be not to sing along with him. I come ashore and see two men and a boy near a stage putting capelin out to dry, while the captain walks on the beach, complaining about how hard times are, how scarce the fish are, and calling on the government for assistance. I leave him to go and look at the cliffs along the shore, and as I walk, I think about how today, one of the boats from the *Clorinde* sent on errand had to wait a short time for an officer; in six minutes precisely, the men using hand lines caught seventeen cod. Where does the truth lie? The fifteen or twenty families living in L'Anse-à-bois manage not only to earn a living from the fishery but to obtain by trading all the items they need. Could it be that they are more economical than the French, have fewer needs to satisfy, and are content with less profit, with what would be insufficient to pay Frenchmen, in spite of the bonuses paid by the government? Every man is entitled to try and obtain the highest pay for the least amount of work. However, the balance between supply and demand places a limit on his ambition and proves to him, by putting a stop to his purchases or eliminating his profits while others may continue to receive them, that the remuneration he demands is higher than the true value of the work done, or that the total amount of work carried out is greater than the remuneration that can possibly result from it. Facts have no feelings. In both cases, the wrong path is being followed, and to persist would be foolish. Fishing must be stopped or salaries must be decreased, or alternatively, if salaries are to be maintained, then productivity must increase by some method or other, for example, by improving the equipment. I fear that our fishermen's efforts, instead of reflecting determination, will consist mostly of complaining.

I return to Mauve Bay via L'Anse-à-bois. I first pass the cemetery, which is surrounded by a fence and a step or two from the blue beach, where small waves wash up. I enjoy communing with the departed: they have always taught me wisdom, bravery, and hope. The graves are dug in the sand, at their foot an upright board with its top rounded in a semicircle and painted white, at their head a marble headstone similar to the board but much larger. They all bear an emblem: two hands interlocked, a hand reaching out of a cloud and picking a flower, or a hand offering a rose. These images of a bond in death after a bond in life, of the will from above to pick a flower down here, of the resignation of someone to give his life back without a murmur to the

One who gave it to him, all speak of the virtues of those who drew them. The art is crude because art is an expression whose meaning is scarcely grasped by the English. On the other hand, it proclaims those virtues which are very English: family, faith, belief in the hereafter, and courageous submission. Then, on some of the stones, you read the message *Gone Home* or a phrase from the Bible. Below are the names and ages of the deceased. There are men and women; one was forty-five, another poor old lady was eighty-four and was buried with *her beloved son* and her grandson. The poor as well as the rich need room to live. Scotland, her native land, was full, so she left; the world is wide, and so she found a new land to make her home; there she was able to live, carry out the duties of raising a family, the joys and pain that are the lot of every human being. She died there in old age, enjoying the ultimate happiness of being able to mix her own dust with that of the ones she loved: *in love they lived, in love they died,* as the old ballad goes.[7] There are also children buried here, their small graves covered with fine sand; pious hands have decorated them with a wreath of white beach rocks. In the middle, set into the ground, there is a pane of glass. I do not understand why. Is it so that the mother or father will be able to have the horrible joy of contemplating the wood that on one ill-fated day locked away so much tenderness, anguish, love, and useless devotion? I do not know; it is impossible to see anything; the glass is misted over underneath. The earth takes pity and seems to hide the coffin under a veil of tears.

In French colonies, the cemeteries contain no women's or children's graves, only those of men, aged from twenty to thirty. This has serious implications.

A little farther, at the top of a hill, is the church, built of wood. The windows are large and the inside is clean and bare; at one end is a chair with a Bible resting on it. Opposite it, there are two parallel rows of benches: a Puritan church for poor Puritans, with none of the trappings that so often distinguish our churches from those in the country and that make what could be so moving into something grotesque. Lower down, the few houses making up the village are scattered haphazardly, since the land has no value and the English like their independence. Many of them are built near the sea and have an attached wharf extending out into the water. The inhabitants are busily working. Tall, blond, and strong, the men are stacking fish in barrels or going out to fish; the women

Newfoundland (Northern Peninsula). Sacred Bay [Album Rock]

look after the houses and from time to time appear on their doorsteps to look at the children playing in the grass. The boys slide down the grassy slope of the hill and shout, while the girls, dressed in white aprons with bibs, are more serious and calm. One small, bareheaded boy has lost his shoe; his plump hands, red from the cool wind, are crossed behind his back, and he is intently studying a fly as it walks around a dried capelin in the sand. Everywhere there is work being done; everywhere there are families and nowhere is there a single uniform. Philosophers and historians, you who study the lives of nations and try to understand why a certain race populates the earth and why another is powerful and why still another slowly diminishes in number and disappears, instead of studying and counting numbers, reading and writing volumes, trying to go back to the beginning of time, why do you not walk for a moment with me through the village of L'Anse-à-bois, Jacques Cartier Bay in northern Newfoundland!

Sacred Bay is not very beautiful to look at; it is a wide expanse of water without character, bordered by hills of medium height, covered with endless dwarf fir trees. Mosquitoes and blackflies seem to have chosen it as their headquarters; seldom have they been worse than this: every boat that comes ashore is followed during the return trip out by a cloud of the insects, which fiercely and unrelentingly attack the faces of the sailors, who cannot take their hands from the oars to drive them off. At the opening between Cape Onion and Corbeau Point lie the Sacred Islands, barren rocks along which a beautiful iceberg is passing at this moment, and a small island, a result of erosion, flat and barely above sea level, whose central portion in the shape of the base of a cone makes it look like a old, floating hat. The interior of the bay is dotted with isolated rocks, one of which is named the Mauvais Gars Rock.[8] What vessel must have encountered problems there, that its crew, unable to do anything better or worse, took revenge by giving this insulting name? At the head of the bay, there is a steeply rising cliff called Album Rock, a reminder of an episode that involved Admiral Cloué,[9] who devoted a considerable part of his naval career to charting the waters around Newfoundland, during one of his stops here. The admiral had collected a number of picturesque views of the country and had the idea, for the title page of his album, to have the word *Album* written in huge letters on this cliff and to photograph it.[10]

On the beach, a fine vein of iron pyrite runs alternately into and out of the water; the mineral rock has disintegrated, and the sand is mixed with crystalline balls as shiny as gold or ochrered, some of which are the size of a fist. There are also erratic blocks in protected spots, in the shelter of a point or promontory; they are rare and quite round, pink granite, finely layered white gneiss; sometimes you see syenite and red sandstone from Labrador; in open areas, these are extremely abundant and belong to the adjacent terrain. When they have just become detached from the rock, they are angular; then they are carried toward the sea; each winter they are lifted and moved a certain distance by the ice, then deposited when it thaws, only to be picked up again the following winter until they become trapped in a strong enough body of ice; they are finally taken by the current from the north and travel to the Gulf of St Lawrence to pile up on the seafloor, raising its height, or are washed up on the southeast coast of Canada or the west coast of Newfoundland. Sacred Bay is a minor destination but a major departure point for erratic blocks. This is an example of nature in the process of carrying out one of geology's most important phenomena, the rapid transformation of underground and underwater terrain.

One morning I go to examine the area around Fauvette Point[11] – a very pretty name for a very sad place – and I visit the location where a cabin once stood, where there are now only four small mounds, the last vestiges of a wall overgrown with vegetation. European civilization is indeed disappearing more and more from Newfoundland; every year, fishing settlements that were once prosperous are abandoned. What has become of the people who lived here? The grass grows and the rain erases all!

Back on board, I notice that the coastline changes shape before my very eyes; it rises above the horizon to about three times its original height; at the base appears a stem which becomes narrower and narrower while the upper portion opens like a mushroom. You would think it was a tall mountain bisected by a layer of water. The appearance is bizarre; near a lake, the reflected image shows the object in reverse. Both are symmetrical, but because of the mirage, the two halves are different. A boat on the horizon sometimes produces three superimposed images separated by transparent bands. This effect is quite frequent in these waters, and it is invariably a sign of rain and bad weather to come.

The forecast is right, which is not always the case. For three whole days, the heavens open up and the ship is under a deluge. The wind blows violently; the angry sea makes its swell felt everywhere in the bay and rolls the *Clorinde* with her two anchors in the water; the low clouds run frantically across the sky and through the grey curtain; we can see the *Drac,* the Division's escort vessel, which has come to join us and share in our misfortune. Boredom reigns on board; it is an impression that must be experienced in order to be fully understood. The hours drift slowly and monotonously by, the wind in the rigging sings the same note, one roll of the ship follows another identical roll, as though it were a huge pendulum, accompanied by a dull, prolonged creaking of the entire structure. You feel too miserable to work, to write, or to think; you try to converse, but the answers follow one another languidly and at ever-increasing intervals until they stop altogether. Everyone reclines in the mess, as comfortably as possible, with a book in hand. Presently, the book droops, rises again, droops again in sequence with the reader's eyelids, which close, open, then close again. When the book falls, his eyes are completely closed and the sleeper's breathing blends with the ship's groaning. The officers are lying in a circle, with one's head almost touching the next one's feet. The only one with the talent to find a last burst of energy is my friend K.,[12] who picks up his album and pencil and cleverly sketches the scene in front of him and entitles it "Dejection." You raise yourself up on one elbow, smile, and then collapse again after such a terrible effort. Occasionally, you go up on deck where the men on watch walk in the shelter of the canvas, dressed in their oilskins and sou'westers. Since there is nothing jolly about all this, you soon go back below. Mealtimes don't bring back any life to the mess; when you lie around all day, you have no appetite when dinnertime comes and then suppertime, and you do not feel sleepy in the evening. However, you try to eat, you go to bed early, not so much in the hope of sleeping as from a desire for a change of scenery, to see whether the air in the cabin possesses more inspiring air than the mess, and, after all, you do have to go to bed. Night is longer and even more tiresome than day. You spend hours observing the glow of the lantern as it filters through the openings in the wall and draws a line across the ceiling; you hear one round following another, you listen to a sailor snoring in his hammock and envy him, you doze off and when you reawaken, you grasp your watch, only to learn with dismay that you have hardly slept more than a few moments. You start all over again, looking at the glow of the lantern, counting the rounds and listening to the sailor. How welcome it is to hear the reveille sounding gaily, the bugle and drum at five in the morning. You quickly rise and go on deck, full of good intentions, intending to start work and to accomplish something. But, alas, the rain is still falling, the wind still blowing, the sky still grey, and today passes just as sadly as yesterday.

Finally the bad weather improves and morale rises with the barometer. To finish getting us back on our feet again, the *Lily*, an escort from the English squadron en route to St John's, enters the bay and graciously offers to wait while we finish writing our letters and to transport them for us. We get to work, and though it will be difficult to produce literature, we manage to write letters that will be as pleasant for those who are at home as the ones received from home are to the traveller, although they contain only the words, "I am well and I think often of you." We return to our cabins, write, then the letters are carried to the *Lily*, which pulls up anchor immediately and prepares a salute; our six marines dressed in their red uniforms stand at attention on the bridge, the bugle plays a rigadoon, and she is gone. She disappears around Corbeau Point, leaving behind a trail of smoke; leaning on the scuttles, we watch her until she is out of sight.

The weather has improved completely. The next day we take the steam launch to Ha-Ha Bay, where, from the top of a hill, we are able to survey a small lake between two bays. We pass Cape Onion, with its cliffs and sea caves, and land in Ha-Ha Bay on a beach of schist pebbles, most of which are covered in a layer of white limestone secreted by the bryozoans that live at the bottom. These animals grow so rapidly that in their centre, consisting of something that resembles chalk, I find mussels. Although most of these are dead, several, in spite of having their growth forcibly stopped, are still alive, surrounded on all sides by four or five centimetres of rock. We climb the hill and, from the top, have a view of the surrounding country. Our lake is shining in the sun like a brilliant spot amidst the greenery. The region is flat; water gathers in the lowest parts and the shape of the lakes varies depending upon the amount of rainfall that year. Once the operation is complete, we climb down and on our return journey meet an iceberg that has turned over so many times since it left Greenland that its surface is carved with parallel grooves in cylindrical rolls from one end to the other. It is reaching the end of its travels; every minute another piece breaks off it, a tiny parcel, a toy iceberg, which the sea rocks while waiting to devour it completely.

We arrive back on board just in time to witness the only hunt of our entire campaign in Newfoundland. God knows we are not short of guns, ammunition, and instruments of destruction, each one more efficient than the other. We even have hunters, which is rare, who are determined to massacre the animals that we are told abound on the island: caribou, with their magnificent antlers, black, blue, silver, and red fox, beaver, bear, the ferocious white polar bear, who has come ashore from an iceberg. In fact, we would be content with a muskrat. They do not see a single thing, our poor hunters, although they spare no pains: they run, trot, and splash, and conscientiously let themselves be devoured by mosquitoes, and the most fortunate only escape the shame of returning empty-handed thanks to a couple of miserable robins. Another time, there is a sheep hunt – for tame sheep from St-Pierre and Miquelon that had been on board with us for two months, wandering on the deck, dirty, pitiful, and wet. We had put them ashore on a tiny island to give them a chance to get their health back by grazing on fresh grass. The island is an island only at high tide; at low tide it becomes a peninsula. The sheep crossed the isthmus, and once they were in the bush, the most inviting words were no more successful than the most frenetic chase. Since we cannot resolve ourselves to leave behind six legs of mutton, a number of chops, and all the rest, three bullets soon get the better of the three sheep. So goes the story of our hunt, and I cannot be blamed if it is not more interesting. On the other hand, while game is scarce, fish are plentiful. Each meal is a veritable icythological orgy: fried capelin, trout in *court-bouillon*, cod in white sauce, plaice *au gratin*, lobster, winkles, and, finally, grilled salmon. At first, it is exquisite; then we become rather more difficult, and the dishes circulate without anyone touching them; fresh cod is the least popular; then, one by one, we scorn them all, even the trout and salmon, which are the best, even though they are incomparably inferior in taste to their European counterparts. Yes, you do become satiated with wealth, as the poets say about riches, and I confirm it in the case of fish, with which I am more familiar.

9

The Cod Fishery

Newfoundland, along with Iceland, is the land of cod. Without cod, I do not know whether Iceland would have been discovered as long ago as it was, because those hardy Norse sailors plied the vast ocean aboard their *drakkars*,[1] braving storms, to discover Greenland and Vinland, that mysterious part of the United States.[2] Fleeing from famine that brought the wolves out of the forest, they went to seek their fortune, but especially to find food.[3] It is certain that without codfish, Newfoundland would have remained uninhabited for a long time.[4] Its coasts are guarded by a belt of ice for part of the year and during the remainder by fog, rain, and gales; the climate is inclement, the cold lasts seven months a year, and the short summer which follows, when the sun is almost always covered by cloud, is unsuitable for any grain to ripen, with the exception of barley. The harvest is precarious, even for potatoes. The soil is poor, from the east to the west; everywhere there are lakes, marshes, and huge forests of fir trees, growing in damp moss, with none of the great rivers by which life penetrates into the heartland of a country. There are a hundred other places around the globe that would be more generous to man, but Newfoundland offers those almost inexhaustible resources of the sea, seals, lobsters, and especially cod. That is why Newfoundland was populated, why it has been fought over for so long, why the terms of a treaty more than a century and a half old are still causing such interminable quarrels over there, fisticuffs and blows for the fishermen and in Europe so many accords drafted by the diplomats.

The ocean is full of living creatures. In animals and in mankind, historical phenomena follow the same laws. Population migration is from north to south. The ocean currents that descend from the Baffin Sea carry great quantities of infinitely small organisms, which larger animals such as herring, capelin, and squid feed upon, and in turn these are hunted and devoured by larger species such as cod, which are hunted by even larger species, fishermen, who catch them, salt them, eat their flesh, and discard in the ocean their heads and entrails, which serve to nourish small fish and so complete the eternal cycle. In France, even the economy of certain industries is affected; in fact, it has been claimed that depending on the direction in which the wind blows, on the floating debris that constitutes their food supply, the sardines fished along the coast from Brest to Bayonne will be more or less abundant and will be found near the land or not. In Newfoundland, when there are cod, every bay becomes the gathering place for a certain number of fishermen; they repair their stages, fish, split, clean, and dry the cod, load and unload it; there is activity everywhere and you see only haggard but happy faces. When there are no cod to be found, the fishing settlements are abandoned, the stages fall to ruin and collapse and rot in the grass; no more boats moored, no dories in the harbours, no life anywhere. The only thing a traveller sees, which is always so heartbreaking, is either solitude and abandonment, or misery for the remaining sedentary fishermen, who try to outlast their bad luck and obstinately hope for prosperous times.

Since the Treaty of Utrecht, the cod fishery has been regulated in the following way: France has the right to fish for six months of the year, from April 5th until October 5th along the coast known as the French Shore of Newfoundland, from Cape St John north to Cape Ray, while the English have exclusive rights to fish from Cape St John south to Cape Ray. The fishery on the banks is unregulated; our colony of St- Pierre and Miquelon has such good fortune because those banks are so close.[5]

The fishing boats are equipped differently, depending upon whether they fish on the French Shore or on the banks. For the former, the coast is divided into a certain number of precisely delineated fishing grounds, each with a name and a stage of very rudimentary construction since it must not be permanent in nature. Stages are carefully repaired every year at the beginning of the season, and several cabins, where the crews and workmen sleep, are built of wood like those we saw at Jacques Cartier Island. One bay may contain several fishing grounds. Every five years, a lottery is conducted in St-Malo under the auspices of a naval commissioner, who assigns one fishing ground to each ship owner for a period of five years; it is possible for an owner to make a friendly arrangement with his competitors and exchange one ground for another that seems likely to be more advantageous and to allow his boats to work better together. As for the ships that fish on the banks, *bankers*[6] as they are called, they are outfitted in France or in St-Pierre. There are distinctions made between the ships from France according to whether they intend to fish and salt their catch on board or fish, salt on board, and also dry the fish on land. All the ships that fish along the coast dry their catch on land. Finally, a great number of small, hundred-ton schooners manned by crews of sixteen have been built in St-Pierre; they winter in the colony, lined up tightly side by side in the *barachois*, and make several trips each season, returning each time to St-Pierre, where they unload their fish, which are then dried on the ship owner's beach. In each of these cases, a subsidy of varying amounts is paid by the government under certain conditions.

The banks extend parallel to the south coast of Newfoundland from Cape Race to the Cabot Strait and are called the Grand Banks, Green Bank, and St Pierre Bank. Banquereau is located on the other side of the huge saltwater river; the depth of the water is between fifty and one hundred metres. The banks are formed by solid matter carried and deposited there by the currents from the Gulf of St Lawrence and especially by coastal ice that breaks away from the west coast of Newfoundland each year and melts when it comes into contact with the warm water of the Gulf Stream. That is where the cod congregate with other species, none of which is edible except halibut, which is salted but is not worth as much as cod. The largest cod are found on the Grand Banks; those fished on the St Pierre Bank are much smaller.

Let us consider the more complicated case of ships leaving France to fish on the banks without the possibility of drying their catch on board. They generally belong to one of the northern seaports, such as Dieppe, Fécamp, or Dunkirk, and head directly for St-Pierre to take on the bait they need for their fishing lines. The species used depends upon the season: in April, May, and June, herring is used; in June and July, capelin; in July, August, and September, squid. The herring, which is larger and of inferior quality

Newfoundland (Northern Peninsula). Capelin drying,
Jacques Cartier Island

compared to the European variety, migrates from the north and
is fished off Newfoundland long before it reaches St-Pierre. It is
fished by the English, who use seine nets aboard fast boats called
gallopers because of their speed, and is taken to St-Pierre, where
it is sold immediately. It is true that this fish eventually swims to
St-Pierre on its own, but if we were unable to obtain it earlier, we
would be forced to spend a month and a half idle out of the total
five months of the fishing season. For this reason, each time the
Newfoundland Assembly has wanted to hinder our fishermen –
we must give them credit where it is due: they have never failed in
this endeavour – they invariably forbid their fisherman to sell
bait. The English may lose in the bargain, but our fishermen lose
a lot more; fortunately, we only depend upon them for herring.
Capelin, which is as big as a sardine, arrives in St-Pierre from June
12th to 15th in such vast quantities that it is almost terrifying: the
entire shore is covered with them, just as our coasts are covered
with seaweed after a storm. You could fish them standing on
shore with a shovel; however, it is better to go knee deep in the
water and use a basket. Finally there is the squid, which is a small
cuttlefish that is caught using a jigger, a type of cylinder made of

lead painted red with a row of hooks around the lower end; all you do is move it up and down in the water on the end of a hand line. Once the bait has been taken on board, the ship heads to the Grand Banks, chooses its spot, anchors, and begins to fish. In the past, large trawls pulled by three and often by as many as eight men were used; now, all you see are dories, flat-bottomed boats with no keel, which were invented by the Americans and which we had the good sense to borrow. They are light, and can be stacked evenly on top of one another, which makes them easy to transport, since six or eight dories hardly take up any more space than one. They can be hoisted by two men, which is quite a saving, because if one of them is lost, only two men perish and the fishing is not slowed down as much as if an eight-man launch were lost. Dories are so seaworthy that they have been reportedly found afloat after a storm still intact, with the bodies of the sailors dead from exposure or hunger still on board.

Formerly, seines were used to fish; this technique is still in use a little in the north of the island; on the banks, only hand lines are used and especially bultows,[7] which are long cords with shorter cords attached at regular intervals with a sturdy hook on each one; they resemble a Mediterranean *palangre* or long line. The hooks are baited and the lines coiled and placed in buckets. At four o'clock in the evening, you go out in a dory, equipped with a compass and a conch, a large seashell with a hole; if you are lost in the fog, you blow into it and it produces a sound that can be heard at a great distance. So, if the fog lifts, you have a chance to find your way back to the ship. A small anchor or grapnel is thrown overboard with a rope tied to a barrel with a pole and a piece of cloth, and the line is let out gradually as the dory is slowly rowed away. When the whole line has been set out, it is attached to a second barrel similar to the first, and you go back to the ship. The banker is like the hub of a wheel, and the lines are its spokes. The next morning at four o'clock, the dories go out again, the lines are hauled, the fish are removed from the hooks and thrown into the bottom of the boat, and you go back to the ship; the fishing is done and the preparation begins.

The cod is prepared in the same way on board the bankers as it is on land. The splitter[8] sits in front of a stall on a wooden chair or stands in front of a barrel sawn in half with its back split open to permit easy access, armed with a pointed, narrow-bladed knife, his hands protected from the sharp fish bones by gloves. He picks up each fish by the eyes with a single movement, slits it from the throat to the belly, removes and discards the guts, throws the liver into a bucket nearby and the eggs into another, and finally removes the head, but without cutting it, in order to preserve the *chignon*,[9] the most elegant part of the cod, which is shaped like a pointed shield. He passes the cod to the splitter, who with his left hand covered with a heavy canvas mitten protected with leather, holding a rectangular, double-edged, wide-bladed knife, splits the fish from belly to tail by cutting halfway through the backbone. He then sends it along to be salted. The salter[10] is down in the hold, where he spreads the cod on a layer of salt; when he has completed one row, he covers it with salt and begins another row. The fish prepared in this way is called "green." The eggs are sold to sardine fishermen, the tongues are set aside, and the livers are put into large barrels kept at the stern of the boat, where they decompose, which allows the oil to separate. The oil is later decanted and collected while producing the most abominable stench imaginable.

Once it is fully loaded, the banker either heads back to France or, more frequently, to St-Pierre, where it transfers the green fish to a faster, oceangoing ship that will carry it back to France as quickly as it can to prevent the fish from being spoiled by the heat and losing half its value by turning into what is called "red cod." France is where it is dried, while the banker returns to the banks and resumes fishing.

The St-Pierre schooners fish the same way, but because they are smaller, they require less time to load and return more frequently to the colony; instead of entrusting their catch to the larger, faster ships, they deliver it to the ship's owner, take on another load of salt, and make sail for the banks once again.

Upon arrival, the cod is passed to men who stand knee-deep in water near the dwellings, warehouses, and beaches and wash the fish one by one with a brush to remove the salt. The fish is drained and spread out on the large rocks of the *grave* or beach to dry, laid in even rows where there is not a single blade of grass. The air is able to circulate below as well as above the fish. The drying process continues until the cod can be picked up by the tail without bending; the drying is then complete and the cod can be kept almost indefinitely. There are other techniques and devices for drying cod, such as flakes or racks that can be inclined toward the sun at different angles, which we have already described but which are used much less frequently than *graves*.

On the island of Newfoundland, the fishing is carried out differently. The ship arrives at its fishing ground loaded with salt; the crew catches its own bait or purchases it from one of the many English schooners that sail the coast. Cod are fished using hand lines, seines, or traps. Usually, each evening, the men return to the fishing stage. There, using a pue – a long wooden handle with a sharp prong or tine attached[11] – they pick up the cod by sticking them in the head to avoid damaging the flesh and toss them onto the wharf, where they are prepared. As be-

Newfoundland (Northern Peninsula). Cod liver oil barrels, Jacques Cartier Island

fore, they are gutted, split, and salted. As soon as they are well salted, they are washed and then dried on the beach. At the end of the season, the tents that shelter the stage are taken down and entrusted to the care of an Englishman, the cod are loaded on board, and the ship returns to France.

On the west coast, the fishermen follow the cod, which follow the capelin as they try to escape by swimming north along the Gulf of St Lawrence; the fishermen "run the Gulf," as the expression goes. They stop wherever the fishing is good and move on if the fishing is no longer worth staying for. Thus, the length of time spent in each fishing ground varies. When the hold is full, or there is a risk that the green fish will not keep, it is unloaded at a certain point along the coast, such as New Ferolle or Old Ferolle, near the entrance to St Margaret Bay or Port au Choix, and dried on land.

Finally, there is the method used by the inhabitants of Miquelon and L'Île aux Marins; they fish inshore in boats called *warys*[12] or dories with two men aboard and sometimes a cabin boy. These boats do not require a substantial investment. Every evening, the fishermen return to shore, the fish are quickly unloaded while women and men alike wait impatiently, and everyone sets about gutting and splitting. I have seen a few of the sites they use on the isthmus of Langlade. The installation is limited to what is strictly necessary: a shelter to sleep in, a storehouse for the salt, a few planks laid on trestles for an outdoor stage. I once stopped for a moment to watch the people at work, and my presence was scarcely noticed because nobody had a minute to lose. The gutted cod pile up and the heads litter the beach; fortunately, the waves soon take them away. The cod are fished using hand lines baited with live cockles, which are bountiful on the sand from the Barachois to Langlade. When the cod are satiated on capelin and no longer attracted to it as bait, they use a *faux* [jigger], a sharp, double-sided lure or hook attached to a slender metal fish some fifteen centimetres long with a hole at its tail to tie on the line. The fisherman makes a swinging movement of the line, and the hooks cut into the school of cod; when he feels a resistance, he pulls the line in. It is a barbaric custom that wounds or kills twenty cod for every one that is landed.

Several industries have developed around the cod fishery, such as the preparation of cod livers and cod liver oil for medicinal purposes and the construction of boxes and barrels. The livers are always prepared the same day they are caught, and cleaned of all impurities; the oil is decanted very frequently to prevent it from going rancid. It is amber in colour and tastes less disgusting. Many other industries could be set up in St-Pierre; among those which have been tried are collecting cod offal and transforming it into manure, and manufacturing salt by freezing seawater. Another idea has been to dry codfish in artificially heated vats, thus saving part of the cost of labour, since fuel is inexpensive and the operation would be better controlled and more regular. This would avoid the risks of drying the codfish in the open air. It would also help improve the quality of French salt cod and allow it to better compete with English cod. In France, the subsidies are so high that cost of cod is prohibitive, at least in foreign markets and in the colonies, which are the main customers. Although these projects were expected to be profitable, they all failed. Routine is sacred. France is the land of inventors but not the land of invention. There is not a single invention that has met with a favourable welcome, unless it has first taken a detour and returned to France via a foreign country, in which case it has a slight chance of succeeding. It is enough to make you believe that the spirit of invention is like love: it lives on air alone and dies if it is fed.

The cod fishermen belong to several categories. Those who arrive from France aboard bankers, and those on the Newfoundland coast who, like all sailors, are hired for the season or for the duration of the voyage and are paid by the month, never leave their ship. Upon their return, they receive a bonus in proportion to the number of cod caught. The population of St-Pierre and Miquelon is sufficient to man the local schooners; if it were not, the ship owners would send sailors from Metropolitan France, and return them at the end of the season.

At the present time, salaries are regulated. Generally, a sailor remains in the service of the same ship owner. He has a notebook in which everything he or his family receives on credit during the winter is written down during the season he is employed; in this way, he avoids having to handle small sums of cash.

The final category of men come from France at their own risk and peril. They try to obtain employment either on a boat or on land, or they hire themselves out to the inshore fishery. If they wish to winter in St-Pierre and avoid the cost of a return voyage to

France, they must obtain permission from the Maritime Registry and pay a fee.

Once a schooner is back in the *barachois* and all its gear has been stored and the cod have been delivered, the value of the catch is calculated based upon the average price for cod during the season. The total is shared in three equal portions: two for the owner and one for the crew. This portion is then further divided: one or two shares for the captain of the schooner, a one-and-one-quarter share for the first mate, a single share for the *matelot*, or sailor; two-thirds or three-quarters of one share for the *novice*, and half a share for the cabin boy. The ship's owner also collects a bonus on two-thirds of the total catch, usually agreed upon in advance. The supplier pays himself what he is owed based upon the books; if the amount is contested, these books must be examined by a judge and the difference paid to the sailor. Each sailor can earn from 800 to 1400 francs for a voyage.

The fishing world is a curious one to observe. A history of the fishery, going back if possible to the first ships that set out to fish for cod, especially under Louis XIV and Louis XV,[13] during the height of the French Navy's power, would be extremely colourful and interesting if written by a man of intelligence and wit. But of course, there is no topic that a man of intelligence and wit cannot make colourful and interesting. France's influence throughout all of Europe also spread to North America, through Canada, and as far as the shores of Newfoundland; the language and customs are the last traces left behind by Champlain, Frontenac, Cavalier de la Salle, Jolliet, Lemoyne d'Iberville, Céleron, Montcalm,[14] and so many other explorers, generals, and governors whose talent, dedication, and courage are now lost to us. The traveller feels surprised and wistful when he hears the words, expressions, and turns of phrase used by Marinette and Gros-René[15] that are more familiar to his educated reader's eye than to the ear. However, everything is eventually modernized. I have seen Andalusian women replace their beautiful *mantille* by a hat with flowers *a la ultima moda de Paris* – the latest fashion from Paris, but no more graceful. The Turks still wear their frock coats and their country is no better off; the Japanese wear shiny boots and are no more majestic. The French language in Newfoundland is disappearing day by day.[16] It is an impoverished language, closely related to Canadian French, as a farmer is related to a fisherman; it is hearty and colourful, with a frankness that can sometimes recount an entire history with a single word. *Boëtte* – bait – is so called because the sailors who served under Duquesne, Jean-Bart, de Tourville, or Duguay-Trouin, Monsieur Duguay of Saint Malo,[17] who fished in Placentia Bay with their loaded muskets in their boats, heard some English prisoner refer to it as bait, from the English verb *to bite*.[18] The words *grave* [beach or bawn] and *chauffaud* [stage] are Breton and Norman. How colourful the word *piquois* [pike, pue] is for the tool used to gaff the fish [*piquer,* in French]. Its sound is quick, clear, and sharp like the instrument it describes, and whoever has seen one understands that no other word would do. Fir tree trunks placed side by side to form the walls of a cabin or a stage are called *orgages* because they resemble the pipes of an organ. What poetry there is in the word *brihat,* used to refer to the streams in Newfoundland, narrow torrents that flow noisily under the vegetation, jumping from rock to rock. What lovely old language, what lovely old times! I know it is not fashionable now to speak of them in glowing terms, but not all was bad in the past, and today not all is good.

A cod fisherman's trade is a hard life; you can hardly imagine all the worries, difficulties, suffering, danger, sickness, injury, and death that it takes to produce a meal of cod, which is sometimes eaten almost with distaste. The misery is widespread; all fishermen suffer, from the ones that fish along the shore to the ones on the banks, from the cabin boy to the ship's owner. The former are more to be pitied; they risk their lives, the latter risk only their money. The difficulties are delayed, from the time the ship leaves its home port to the time the dried cod are securely stored.

As soon as his boat arrives on the banks, the fisherman works continually. At dawn, he hauls his lines; once he is back on board he must untangle them, because they are often tangled up by fish trying to escape; he baits them and must immediately set them in the water again. Not until ten or eleven o'clock in the evening can he stretch out on the straw in his bunk, where there is almost no air and what little there is, is ill-smelling; he lies in this coffin with his clothes still on, and takes a few moments of rest. Most of the sailors do not remove their boots during the entire season. They are in the water continually, drenched by the waves and the rain, with the horrible odours of the putrefying livers, the fish entrails covering the water in a vast oily layer, and the hold full of green salt cod. The rest of the time might be peaceful, but the ship is held at anchor and tossed by the waves; if the wind freshens, if the waves increase, preparations must be made to let out more cable or to cut it if absolutely necessary. When there are plenty of cod, the piker, the splitter, and the salter all need a helping hand, for the fish cannot wait. Dampness produces wounds that quickly infect; what remedies there are do not work; the only true one would be to stay dry and to stop work, both of which are impossible. The infection spreads and worsens, until the finger, the hand, or the arm must be amputated; those who are able to withstand the operation are the fortunate ones. "That's what it takes to do the job," the poor men say stoically, and that is the extent of their complaining.

There are more serious dangers. The fishing banks lie on the route used by the huge ocean liners that travel back and forth between France and Ireland and Newfoundland, Halifax, and New York. Several of them pass by each day. In thick fog, it is impossible to see more than a hundred metres; furthermore, the steamships are racing one another at terrifying speeds. They steam ahead with all lights blazing and the whistle blowing constantly, with the bells ringing and the foghorn bellowing; but these precautions are often useless – the sound is muffled by the fog and the sea. The fishing boat on the banks blows its horn and rings its bell, but being anchored, cannot manoeuvre, and the steamer's sharp stem cuts through it; there is a slight shock followed by a few shouts, then silence as the fishing boat sinks straight to the bottom. Twenty, thirty, or forty men die, and the liner continues on its way, often not even signalling the accident, and if it arrives in New York ahead of schedule, everyone will say it was a magnificent crossing. Women and children wait at home in Brittany as the months pass; the widow and orphans live in poverty, survive however they can, selling seaweed or begging until the oldest is old enough to go to sea as a cabin boy, the almost promised victim of a sea that will claim the whole family. In a village in Brittany, I met a woman who had lost her two sons, her husband, her father, and her grandfather in this way; they left and never returned. Her brother was a sailor and had died in Tonkin,[19] and one of his two sons was a cabin boy and planned to go to Newfoundland the next fishing season. While she told me about her loss, as though it were something simple, a terrible fate that could not be altered, the other son, a small child hardly able to walk, played in the sand near a tidal pool on the beach. The sea has taken the lives of countless men, so Breton women have good reason to always wear black.

That is not all: icebergs drift along the east coast of Newfoundland and on the fishing banks; hidden by the fog, they advance slowly and stealthily, carried by the current. And by the time you notice a white glow surrounding the ship, it is too late, the ship is crushed.

The men who fish along the coast are more fortunate; they go ashore every evening and sleep in cabins around the stage; if they fall sick, they are cared for. Every ship carrying a certain number of men must have a doctor. The doctor treats the sick and splits cod. Under such conditions, the medical personnel do not consist of outstanding physicians. Most of them are former sailors with little training, who have passed a simple examination, or students of thirty years who have had many problems, who are disillusioned and spend their winters tending a small garden in France and their summers fishing. Their remedies are simple, like the illnesses they treat. They prescribe a little rest and a lot of cod liver oil because the barrel is nearby and they know no one will overindulge; they purge, they induce vomiting, they bandage wounds and scratches, and as for the patients, may God protect the poor devils.

Once the cod is fished, salted, and taken to St-Pierre, it has still not reached its final destination. If it is too green when it arrives in France, and if all the necessary precautions have not been taken, if the ship transporting it is delayed, the cod turns red; if too much salt has been added, the cod is 'burned' and cannot be sold; if the weather is windy or the sun too hot, the surface of the fish dries too quickly and the evaporation of the moisture in the thicker part of the fish stops and it is unfit for sale. The same monotonous refrain is heard over and over – the wind, the sun, the fog, the rain, spoiled fish, spoiled fish! Oh what a horrible living for all those who work in the cod fishery.

10

The Eastern Shore and Cat Arm

*T*he *Clorinde* heads south to map the waters of Cat Arm, at the mouth of White Bay, opposite Cape Partridge. We pass Pointe des Graux and the question is whether the captain will decide to enter Croque Harbour. While in port, we have a tendency to become sybarites, to prefer the quiet night at anchor; when we get up in the morning we push the porthole open and breathe fresh air as we wash. Moreover, there is a rumour that the vegetables that we planted during our last stop here are probably ripe; such serious news creates considerable emotion. The captain is on the bridge, silently stroking his beard; he alone on board knows whether we will soon be eating radishes. What a beautiful thing to command a ship! The frigate sails on and our hopes diminish; suddenly, she turns to starboard and heads toward Croque. Hurrah! We are going to eat radishes.

The anchor is no sooner dropped than our Robinson Crusoes, our cattle keepers, are aboard with the long-awaited radishes and some cress, just enough for the table. The bugle calls us to dinner; the *maître d'hôtel* himself respectfully places the fresh green vegetables, which are a delight to our eyes. Without speaking, we raise our forks to our lips, then our teeth sink into the exquisite and crisp freshness; we eat everything: the radish, its tail and the cluster of wrinkled green leaves, which pleasantly tickle our taste buds. As for me, I have five, which I savour one after the other like a gourmet. We finish with the pungent cress. That dinner is a feast to us; let anyone who is astonished by our enthusiasm go two months eating lentils, beans, and preserves and then sit down to a plate of radishes and cress.

Since our visit, an iceberg has gone aground at the head of the bay and may prove to be a problem for us; we promptly fire two shells at it, which do not have the slightest effect. A moment later, we are rowing around it when we hear a rumble inside it, the result of our artillery having disturbed the arrangement of its molecules. We examine it closely; touch its pretty blue veins and bands where the visible structure of the névé grains is made so apparent by the melting water. We try to climb onto it, but it has capsized so many times that its surface is rounded and polished, and we soon give up. One of the midshipmen is braver, more obstinate, and especially younger, and so perseveres. Leaning on his shipmate, he puts one foot on the iceberg and falls flat on his face. I am content to break off a fragment and put into my mouth a piece of this piece of Greenland. Well, I declare, the taste is identical to the ice in the fountain in my garden. We then break off some larger pieces, which we carry back on board. Our cook tries without success to use them to make iced cheese.

The weather is quite dismal; I nonetheless climb Genille between two showers of rain. I can see the stream from a distance but do not have the courage needed to walk to the lake. You don't mind getting your ankles wet going to see something new, but a second trip is not worth getting wet up to one's neck. Also, the mosquitoes have taken refuge in the woods and welcome all hikers far too enthusiastically. All the radishes are gone, so we continue our voyage.

The coast becomes more interesting as we head farther south and the land rises higher. Already by Croque, the wooded mamelons of the north have made way for hills, and now they take on the appearance of real mountains. The cliffs along the shore are formed from thin, even layers of schist and sandstone, reddish in colour and resembling ancient fortifications flanked by huge towers with broken walls. Elsewhere, they recede, and each step is separated from the next by scree covered with vegetation, so that what you see are alternating vertical walls of red and bands of green producing an extraordinarily majestic effect, a scene like those drawn by Gustave Doré.[1]

We enter and leave Cape Rouge harbour while the captain talks to a fishery judge, who has come up quickly in his boat. We stay just long enough to admire the waterfall at the stern of the ship, which drops from about one hundred metres, blending its foam with that of the sea. At the head of the bay lie six brigs at anchor and in a cove a few stages covered with tents, which seem to be sleeping as well. Beyond, the cliffs rise up higher and higher, and at each drop a cascade tumbles into the sea. That evening, we arrive at Canaries Harbour[2] and drop anchor at Gouffre Harbour,[3] near the entrance.

Nearby, at Canaries Harbour, there is a marble deposit that was once quarried, and the site makes a white spot on the side of the mountain covered with fir trees. It can be reached by a fairly good path. All that was done was to dig a shallow layer from the surface. The quarry was soon abandoned because the marble, although of excellent quality, is so cracked and twisted that it would be impossible to use commercially. During the excursion, I am guided by a good Newfoundlander, quite pale and thin and in despair because of the lack of cod, which has all but disappeared from this part of the coast, where everyone's livelihood depends upon it. He confides in me as we walk, and when we arrive back, he invites me into his cabin, where his wife waits with three beautiful, blond, rosy-cheeked children, full of health and, what's more, clean. The father thinks that the future will be very bleak and that he will have to go elsewhere to earn his daily bread; but the children laugh and play, like birds in the trees. A pat on their little round heads, the last bit of tobacco emptied out of my pouch into Dooley's hand, a kind word, and we have become friends. My guide fetches the bottle of wine that he was given the year before by a fishing captain and has been keeping as a precious treasure in case his wife becomes ill; he opens it reverently and we each drink a mouthful, I to the health of my hosts and they to the health of *my own lady and children*. The hour is late and I must depart; a friendly handshake and I am on my way, taking a memory with me and leaving one behind. We have known each other for an hour and will never see each other again.

One of the sights that has struck me the most during the voyage is Cat Arm, an admirably typical fjord. The formation of fjords is presently one of the least disputed questions of geology, so that apart from the beautiful panorama in front of me, I am happy to have an opportunity to study these glacial phenomena, of which Norway, Iceland, Scotland, and Ireland possess so many standard examples. Fjords are found only in northern countries and are located on the sites of what were formerly rivers or some other break in the terrain forming a *thalweg*.4 During the glacial period, which characterized a portion of the Quaternary Era, a thick layer of ice covered most of the northern hemisphere; the river was transformed into a glacier and for a considerable number of years, a river of ice flowed in the valley and eroded its bed, not like flooding when a river rises and overflows its banks but by cutting and carving the sides of the banks, carrying with it the rock debris, pushing it along to form a frontal moraine. Then the temperature rose and the glacier slowly melted and its surface area and thickness decreased. During this time, the entire region experienced a sudden subsidence or sinking; the melting ice deposited its load of rock here and there; the frontal moraine sank into the sea; the worn, striated rock appeared; and the lateral surfaces became the side walls. Finally, when the warmer temperatures replaced the ice with water, the glacier was partially submerged in the sea, forming a fjord. A fjord is therefore characterized by steep side walls close together, by deep water, by striated rocks and boulders perched in high positions, and by a submarine threshold at its entrance with the least inclined or steep part facing out to sea. Every fjord is therefore a sign that the terrain has sunken; and on a geographic map, by studying the indentations of the coast, it is possible to show where glaciers formerly existed. This is true of all of Newfoundland, particularly of the coast from White Bay south to Cape Ray. It is quite noticeable in Notre Dame, Bonavista, Trinity, Conception, Placentia, and Fortune bays, which have numerous small islands, which are the prolongation of the fjords' sides and the last visible vestiges of the mountainsides that contained the glacier.

Cat Arm has all these characteristics. When the ship enters this narrow, ten-kilometre-long corridor, we marvel at the sight of the mountains on either side. The *Clorinde* cuts through the still water with its bowsprit pointed straight toward the rocks. Suddenly, the wall opens and the ship turns to port, then back to starboard, and the walls become still closer together; when the anchor drops we see the head of the fjord, which looks like a beautiful theatre set. It is a circle formed by successive chains of mountains. A waterfall of some three or four hundred metres in height forms a large pool and then disappears in the vegetation; to the left, a second waterfall opens out into a rock basin; above the rapids there floats a veritable plume of "moist dust"; everywhere you can see the foam of smaller cascades. These waters, particularly during the spring thaw, cause large boulders to roll

Newfoundland (Northern Peninsula). Great Cat Arm (west coast of White Bay)

down; their edges are worn round by friction, and they end up slowly accumulating and raising the seafloor in the fjord. There are also numerous boulders at the head of the bay, where they are half submerged and where the spaces between the rocks are filled with a mixture of sand and mud; gradually the area is covered with grass and finally green trees. The river water flows into the fjord in such volume that it makes it almost drinkable; because its density is lower, it causes a strong current on the surface. It originates in lakes where it is heated by the sun, and when you swim there, your upper body feels warm while your legs are in contact with the ice-cold salt water. The fresh water is the colour of weak tea; at a certain depth, it is dark brown because of the dissolved tannin it contains. The south side of the fjord is covered with vegetation; the fir trees rise above one another; the opposite side burned three years ago and is nothing more than a forest of dead trees, with all the carbon washed off by the elements, leaving the bare white trunks.

Accompanied by two sailors, I set out to climb to the top. The path is difficult; a few steps from the water's edge, there is a band of vegetation and dense bushes marking the boundary of the forest fire. It is an incredible tangle of fir trees, some of which are still standing, the others fallen into piles. The rock is visible everywhere because the fire consumed the layer of peat which had accumulated over the centuries and covered the ground; the heat of the fire also shattered large boulders, and the pieces lie spread around. The roots of the fir trees had spread out in the layer of peat but could not penetrate beneath it, and now the charred, twisted roots, held together by the tangle, are visible above the bare ground. A giant could pick up the entire forest lying on top of the rocks, but he could not walk through it.

We climb at an angle to a height of some two hundred fifty metres, avoiding as well as we can the steep, smooth rocks where our feet could slip, and arrive at a crest where a beautiful view opens up below: the calm water of the fjord and the frigate, with only its deck visible to us, resembling a floating nutshell. Yet this tiny wisp has carried us across the ocean and is entrusted with the lives of two hundred people! No one can see us from the ship, although we can hear with marvellous clarity the sounds that rise from her: the bugle, the chiming of the bells, and even the men's voices. We are higher than the mountain on the opposite side; beyond the first crest, there is a second, higher one; beyond that

one, a third one, higher still. At a distance of about twenty kilometres as the crow flies, toward the southwest, a tall, sharp peak
stands out. The waterfall on our left is the end product of a river
which winds around the mountains, disappearing then reappear-
ing farther on and seeming to flow from an elevated lake of which
I glimpse a corner, glittering in the sun. I continue on my way and
suddenly, like a funnel, I see a small lake hidden in the trees.
Using my hands, sliding under the fallen trees, I climb down and
kneel in the thick, damp grass. Dying of thirst, I take a deep drink
of the still water. The moisture helped to protect the area nearby
from the fire; the flames passed over this oasis lost in the middle
of all the desolation. A tiny stream bubbles out as the terrain becomes steeper, it accelerates, rolls from rock to rock, leaps over
those which bar its path, and eventually flows into the sea. Two
robins live here in absolute tranquillity, playing and chasing each
other through the branches.

We follow the crest of the mountain as well as we can; we have
to turn and take a new direction every time a sudden change in
the terrain places an insurmountable obstacle in front of us. In
this way we reach the last perpendicular rocks that altered the
fjord's course. This plateau, which the huge, solid river encountered on its path, bears the marks of its victory in forcing the
glacier to change direction. The bare rock has deep parallel stri-
ations, and large boulders are scattered, balanced in the most unusual positions. Some of them are several cubic metres in size,
others as small as a fist, abandoned wherever the glacier deposited them. "It looks like someone put them there," says one of the
sailors. No, no one has touched them; they have been in the same
position for thousands of years, exposed to the sun, the fog, and
the rain in summer and the snow in winter. This inert mass was
there long before my great-grandfather's great-grandfather was
born, and if I pick it up and place it a little farther on, it will stay
there without moving probably long after the bones of my great-
grandchildren's great-grandchildren have turned to dust. What
thoughts are inspired by comparing the impassiveness of matter
to human activity? Who comes here? How many centuries will
pass before another man sets foot where I have walked? On
mountaintops, the heart feels something of the immensity that
the eye takes in at one glance. I climb down, following a hollow
and making my way through the terrible tangle of dead trees
piled up by the water. The blackflies are ferocious and swarm

around us in clouds; the joy we feel when we arrive back by the sea is indescribable. We are able to refresh our faces and hands, which are covered with soot and burning from fatigue, the heat of the day, and the fly bites.

We have invented a very convenient sort of boat during the voyage. It is made of waterproof canvas stretched over a wooden frame. It has a keel, a mast, a sail, a rudder, two oars and a grapnel; it holds two people comfortably, and because it is very light and folds like a fan, it can be carried quite easily. The only precaution that is needed is to watch out so that a sharp rock does not puncture it. It enables us to sail around on lakes that do not communicate with the sea, even in places where the water is barely knee-deep. In a craft such as this I have made a few trips around Cat Arm, sketching different views of the fjord and, between sketches, walking along the shore and looking into the water at the marvellous world or marine animals and plants: starfish crawling along on the rocks; greenish urchins with five brown double-rows on their backs; jellyfish of all shapes with pink fringes and, inside their bodies, protected by long tentacles, stripes with an occasional beautiful metallic-green gleam tinged with gold. The water is teeming with all these creatures, which move around, fight, and try to stay alive. A crab with large, slender claws and a roundish body, like an enormous spider, hides in a hole and awaits its prey hidden in clumps of grass that undulate in the current. An innocent fish passes nearby, a young one with a small head, *nescio quid nugarum cogitans*;[5] suddenly he finds himself face to face with the crustacean; he makes a sudden retreat and disappears as quick as a flash. A jellyfish smoothly moves the embroidered edge of its circular body, rising and falling in the water; this living jelly is not drifting haphazardly; it is avoiding my approaching hand; therefore, it knows fear and has a memory, since it considers a new object dangerous. This humble creature among the most humble possesses the divine capacity to coordinate facts and deduce consequences from them. Where does instinct end and intelligence begin? What an impossible problem to solve! After all, what difference does the solution make? For nature is uninterrupted, is it not? Does it not form an endless chain connecting all animals, *natura non facit saltum*?[6] All creatures, stones, plants, animals, men, bodies, intelligence, all that exists is interwoven and constitutes an immense network where the thinker, the scientist, can travel at his wish without ever leaving it, starting at any point he chooses, slowing down his steps whenever science provides more refined tools that cannot skip rapidly on to something else. We are prisoners; we can verify the extent of our prison, but does it mean that nothing exists beyond the walls which forever encircle us?

The forests beneath the sea, green and pink algae, some wide and long, others more delicate than the finest lace, bend and palpitate smoothly and softly; the sun sheds bright bands of light on this shivering greenery, and when a cloud covers the sky, dark shadows come over it and then disappear suddenly. You could spend hours leaning over the side of the boat, your head almost in the water, your eyes fixed on the strange scene below where light becomes more luminous and shadows are darker while you contemplate this crystal land and these dull dreams. A stroke of the oar and you are a few metres away, where you discover a new scene, different from the previous one but equally as fantastic. How many times have I envied the lead weight we use in measuring the depth of water when it returns from depths where it has seen things I will never see and that I want to see so badly! I collide with a nest of wild ducks, and the hen quacks and swims away, followed by six ducklings, not yet grown but able to swim like real ducks. We row hard to catch them; every few moments a duckling swims off in the wrong direction and the frightened mother turns her head to watch over her brood and encourage them with her quacking; she swims slowly, pretending to be injured, shows them the way, and as soon as she knows they are safe, she flies off with a victory call and disappears into the brush. All mothers are the same everywhere!

11
Labrador and Cape Breton

We depart Cat Arm; the *Clorinde* sails down the fjord, passes the underwater moraine, and enters White Bay, of which the most easterly point, Cape Partridge, is visible on the horizon, and then we head north. For a few minutes, we see the fault line that indicates the entrance; soon however, it blends with the other lines and forms an uninterrupted row of cliffs. When a remarkable sight has struck our eyes or an event has moved us deeply, we go away feeling as if our heart is broken. But one day follows another, and if we have survived the first shock of pain, when our memory takes us back, we are surprised at how little time and distance our eyes need to forget a sad or a happy moment so close in the past, or our memories to transform it into a pale image that becomes more and more blurred. It is a faint shadow that will be slowly lost in the vast ocean of forgotten memories, where all is destined to be swallowed up, happiness as well as sadness, glory as certainly as love. The march toward annihilation is inevitable; nothing can stop it; a man finds it odious, for he cannot become accustomed to the idea that he will be no more, as though what has once existed will seem never to have existed at all, so to speak; thus he fights with all his strength and protests with every fibre of his being. He carves the word *eternity* on a headstone; nature then continues her work: the rain erases the inscription, the stone turns to dust and blows away in the wind.

I think about these things the following day. There are days when, without knowing why, we are overcome with sadness; we see everything as grey, our thoughts are grey; it is the joys of the trip taking their revenge. I am following the shore of a small bay that in French is called Death's Head Harbour[1] and in English, Maiden Arm, situated in Hare Bay, where the frigate is anchored. There is a story behind this double name.

Many years ago, an English fisherman from the Avalon Peninsula lost his wife, who had borne him a daughter. He wanted to live for the sake of the child, whom he adored, since she was flesh of his own flesh, but much more in memory of the deceased woman. But the sight of the places where he had known happiness was too difficult for him. He left his land, his family, and his friends, took the child, who was his one and only treasure, and headed for northern Newfoundland. When he found a new region and new waters in an uninhabited bay, he stopped and built his hut. The years went by. In summer he fished; in winter he hunted for furs. The bark hut was transformed first into a cabin, then into a comfortable dwelling. Near the wind-swept rocks, the child grew up and became a young woman. She became graceful and beautiful, lived, laughed, and sang; at the sound of that voice so dear, the poor trapper felt his wounds heal and his tears dry. The father and daughter worked and loved each other, and when fishermen occasionally anchored their boats near the small beach, they called it Maiden Bay after she who adorned it. One winter, the girl's beautiful complexion faded, her pink cheeks became pale, her laughter and singing became rarer and rarer, and her father watched her health decline. He felt that she would die; yet at times, he was full of hope, thinking that everything in the world is reborn in spring and that children are not supposed to die. The snow disappeared, summer arrived, the wildflowers opened their petals, the birds, as they had always done before, built their nests, and she breathed her last breath. Between the rocks, her father found a sheltered place where there was a little sand, dug a grave and gently placed his child's body in it. Then he fled. He was never heard of afterwards. In this country, rock appears everywhere on the surface of the ground; the earth is miserly, even for a final resting place that it is forced to provide for the departed. The sand disappeared and when, later, French fishermen came ashore at that spot, they saw a house in ruins and nearby on the beach, a skull, bleached white on a bed of green seaweed: Maiden Arm had become Death's Head Harbour.

The *Clorinde* did not stay long in Hare Bay; we visited Ariege (or Belvy) Bay,[2] where the land is extremely low. Although there are hills visible in the distance, near the sea there are only mounds no more than twenty metres high. The vegetation is dense, and it is impossible to walk through the fir trees. However, in those places where the air and light penetrate, at the mouths of streams, the alluvial deposits are covered with thick grass, dotted at this moment with delightful wildflowers, clusters of violet irises, white daisies, and fragrant hyacinths, a feast for the eyes. In a moment, we have gathered a huge bouquet to bring back on board and give our mess a festive air.

In Baie des Outardes,[3] the mirage appears and soon afterwards the bad weather, so we take refuge in Sacred Bay. By morning, the wind has died down. White clouds are racing along the horizon and the sky is clear once again, though the heavy waves roll; we have entered the Strait of Belle Isle, hugging the Labrador coast.

Labrador, the land of the labourer! What a bizarre idea to give such a colourful Spanish name[4] to such a pale, cold, bare, and deserted land! Names carry with them a certain rhythm, a musical quality that makes us like the objects or the creatures which bear them even before we know them, hate them, or feel indifferent toward them. When Balzac[5] invented one, he felt as though he had completed another act in his immortal human comedy. The combination of syllables in the name Iceland evokes the idea of a dark land, full of sublime horror; by simply pronouncing it, you hear the sound of a block of ice cracking; the name Spitzbergen makes you think of a desolate island, with rugged mountains and sharp peaks reaching up through the fog; when you say Novaya Zemlya,[6] you dream of grey skies and black rocks protruding through the snow. *Terre-Neuve*, Newfoundland, I admit does not convey anything; nothing is more banal than the word *terre*, "land," and in spite of yourself, you hold a grudge against it for wanting to be young and for calling itself new for such a long time.

Labrador looks quite a bit different from Newfoundland. On our left the flat coastline, with no indentations, resembles a black, uneven bar, like the line drawn on the bottom of the quartermaster's carefully written page; on our right, the coast is remarkably high. The land is a vast plateau, very slightly inclined toward the sea, which cuts it abruptly with a line of cliffs. No capes or gulfs can be seen, not the smallest headland anywhere. On both sides, the current has eroded the shore and the debris now makes up the seafloor, from the Grand Banks of Newfoundland to Cape Canaveral, Florida, in the southern United States. The even pile of fallen rocks goes two-thirds of the way up the cliff; its base is continuously eaten away by the waves while it is forever being replenished with fresh, new material, fallen rock, broken off by rain, sun, and especially frost. In the vertical portions there are horizontal, parallel layers of red sandstone; in a few rare instances, there are two rows of fallen rock separating the two layers; it resembles a gigantic stairway, a pedestal with two steps. We pass the Point Amour lighthouse and enter Forteau Harbour, which is wide open to the south and provides rather mediocre shelter when the wind blows from that direction. The bay is a large cut in the plateau. Nowhere is there a tree to be seen, only short, rather thick bushes; at the most easterly point stands the lighthouse; near us flows a river which forms a low but remarkably wide waterfall, three kilometres inland. Then there is a beach with sand of pink quartz, resulting from the disintegration of the sandstone. This is true sand; it forms dunes and rolls under your feet and is a rarity. With the exception of a few coves, I do not remember ever having seen any in Newfoundland. The houses in this land of red are spread along the bay, grouped together near the end of the beach; beyond them, a pretty waterfall drops directly into the foaming waves. The scene is a charming one, thanks to the sunshine which has followed the bad weather of the past few days. The ocean is crisscrossed with pretty boats, narrow at both ends and shaped like whalers, rigged like schooners, easy to manoeuvre and remarkable for their seaworthiness.

We land at or rather climb up onto a wharf which extends from a house, on a primitive ladder connecting two of the pilings. Three boats are tied on here; they have towed their nets full of herring. The fishermen dip spoon-shaped nets into the silvery mass and empty them onto the wharf; the shining, wriggling mountain of fish is loaded into wheelbarrows and stored in a warehouse. Unfortunately, herring do not fetch a very high price; it is of mediocre quality, and they are poorly equipped to salt it. The owner regrets the shortage of cod; all this herring, he says, is better than nothing, but not by much. With a few of the ship's officers, we walk along the shore on a path that is intersected by several streams, which we cross on bridges made of two tree trunks side by side. The clear, transparent water flows over the rocks and does not look at all like the brown, peaty water in Newfoundland.

Labrador. Cod flakes, Forteau

Two outcrops of granite with large, pink feldspar grains form a headland which intersects the curve of the bay. Here the coastal ice was stopped, laden with rocks from northern Labrador, went aground, and dropped its load. Other boulders are scattered farther ahead, only to disappear in the deeper water.

We end our walk near several houses at the mouth of a stream, where a waterfall is just barely discernable in the distance. As usual, the houses have wharves extending from them, where the fishermen are presently busy drying cod that have been brought from the boats that are tied up to the pilings. These people all belong to the same family. The grandfather drives away the dogs; they have short, black fur, a long narrow snout, and pointed ears and jump around in the grass and run up to us, barking. The children look at us with astonishment, and the grandmother asks me whether it is true that this land once belonged to France. Alas, yes! Labrador and Newfoundland, along with Canada, Cape Breton, and part of the United States – all those lands were once French; today, France owns St-Pierre and Miquelon, provided that no fortifications are built and no more than fifty troops are stationed there at any time. The good woman asks why this is so; because the question touches on the philosophy of history and is sensitive and lengthy to explain, I prefer to ask her about the cod fishery. She takes up the same litany we have heard everywhere this year. It is a bad season: people are able to stay alive but no one is earning anything. In these regions, it is impossible to literally die of hunger, because it is always possible to catch enough fish to live on. The land is available for the asking, and seldom is there any difficulty in obtaining a crop of potatoes; hens lay eggs and eat what they find, such as wild berries and even capelin that wash up on the beach. However, if the fishing is not abundant, people are unable to obtain the supplies and various objects that are provided by the merchants or traders in exchange for dried cod and furs.

We enter a dwelling where the doctor has been called to visit a patient, and while he is examining him, I look around the inside of this Labradorian house. It is extremely clean; the dishes are stored on a shelf, table and chairs against the wall; a rug covers the wooden floor, and there are double-sash windows as in all English countries, equipped with blinds in place of shutters. The bedrooms open onto this room, which serves as a common living area, and a few framed engravings are hung on the walls: a portrait of Washington, which makes me wonder about the people's loyalties; the Duke of Orange, proof of their Protestant Irish origins; a *Happy Christmas* and *Good Luck* surrounded by a wreath of flowers with a robin red breast perched in one corner. In the middle, a large iron stove used for cooking throws off an unbearable heat. It is always like this: the temperature may be mild outside, but the stove is filled with wood, and when you enter, you feel as though you will truly suffocate. You have the impression that these people absorb heat so that they are better able to resist the terrible winters. It would be very interesting to talk with them and ask about their lives and customs, but we do not have enough time and must go back on board. The *Clorinde* weighs anchor and we leave Forteau Bay, touching at Port Saunders; we spend the next night moored in Port-à-Port [*sic*] and the following day head for Sydney, Cape Breton Island.

I believe that nobody on the *Clorinde* ever felt such a delightful sensation as the moment the ship entered Sydney Harbour after passing the bare island called St Paul's, sailing around North Point and Niganish Bay[7] on such a beautiful August afternoon. In Newfoundland and in Labrador, the temperature was beginning to cool; now a radiant sun is shedding waves of heat and light. The atmosphere has that transparent quality which on the American continent gives objects extreme clarity of contour and gives those in the distance a particular bluish tinge. The sea is smooth and without a ripple and mirrors the sunlight with almost unbearable brightness; it floods the land without the slightest silver border indicating its edge. What a relief, what a rest for people who for nearly three months have lived in almost complete solitude, to see, smell, and hear life bustling about on all sides! To our left, a semaphore sends signals, to the right, high buildings, chimneys, the huge wheels of a coal mine; farther away, a train's whistle is carried by the breeze and a locomotive races through the trees, leaving a plume of smoke. Unlike the long, monotonous line of fir trees standing in saltwater, whose even tops stretch off like an immense layer of dull green in the valleys and on the mountainsides and beyond, as far as the eye can see, not a road, not a path, not even a miserable hut; we see cheerful green fields, as green as emeralds, which the Ancients[8] claimed could cure melancholy, cottages everywhere, farms with fences painted white around the gardens, scattered amid bouquets of forest. Sailboats pass us continually; steamships, tall-masted schooners, dock at wooden quays where wagons travel back and forth loading coal that you can hear rumbling down into the hold, boats going here and there, a steamer several stories high with its wheels leaving a wide band of foam on the water. The forest still covers vast areas; until recently, the region was wilderness, but man took possession, fought against nature, and was victorious. The climate is warmer, the fir trees are mixed with annual deciduous trees, and the contrast between their different shades of colour is restful to the eye. With binoculars, you can distinguish small, moving black dots in the countryside which are human beings working and cattle grazing, as well as horses enclosed in pastures; occasionally, they panic and gallop around; carriages travel along the roads, which are lined with telegraph poles. This is North Sydney, the mining town, stretching out with the pointed steeples of its churches behind the piers and the masts of the anchored ships. In front of us is the point of land which delineates the bay and where the town of Sydney has been built. It is less commercial but much calmer

Labrador. Wharf, Forteau

than neighbouring North Sydney. At the head of the harbour, a belt of hills forms the horizon of this magnificent panorama. Through a stroke of good fortune, the *Minerve,* the flagship of the North Atlantic, and the escort *Talisman* are anchored facing the town. The *Clorinde* drops anchor near them, two hundred metres from the Burchell dock. At the sound of the chain going down through the hawsepipe, everyone lets out a sigh of happiness. After the bleakness of the Baie du Sacre and Kirpon, after Bonne Bay so picturesque and grandiose nonetheless, we need the rest that awaits us here. Change is an absolute necessity; a man's mind is as rebellious toward prolonged contemplation of the same spectacle as his body is incapable of nourishing itself on the same food continually.

We avail of the first boat going ashore to leave the ship. We disembark on a raised wharf where the steamships dock, climb up a stretch of road, and arrive in town. Built along the shore, Sydney consists of three or four parallel streets intersected at right angles by three or four other shorter ones; this is the same

Cape Breton. Main Street, South Sydney

checkerboard layout of most American cities. Most of the houses are wooden and painted bright colours, most often white, with double sash windows and high, steep, shingled roofs; many have a garden in front separated from the street by an open fence; tall trees shade the lawns and white walls. There are many churches, some of which are built with the brown limestone that is so common in carboniferous terrain and that makes buildings appear old, which is quite pleasing to the eye and the mind. If a house, a factory, or a bridge has the right to look young, it seems that a church, which stands for religion; a courthouse, which symbolizes justice; a palace, which represents the people; a castle, which is a family seat, are almost bound by duty to appear old. In Paris, where the newly quarried limestone is a garish, bright white that hurts the eyes, at one time efforts were made to artificially give buildings an old look, just as we now try to correct it when it happens naturally. Thus things, fashions, and men change: yesterday, today, and tomorrow have different opinions, and often sons and

Cape Breton. Lovers' Pond, South Sydney

Cape Breton. Miray [Mira] River

grandsons like to call something white for the simple reason that their grandfather called it black; if, on the other hand, the grandfather said white, they would prefer to say black.

In any case, the churches in Sydney are clean and well kept. Whether richer or poorer, each one wants to look good to the others; there is a spirit of emulation among them, a sort of honour, for most of our convictions are born of competitiveness. The stores are all grouped on Main Street, the second street parallel to the sea: novelty stores, *dry goods stores*, hardware stores which sell a bit of everything, hotels, barber shops; and, the *drug store* or pharmacy equipped with a soda water machine. There are large multicoloured posters whose upper part shows a vivacious young lady, blond or brunette, with a rosy or a lily-white complexion, wearing a boater's hat, or a portrait of Adelina Patti[9] or Christine Nillson,[10] and the lower part describing the infallible *pain killer*, the effective green, blue, yellow, red, or black pills, the indispensable *liniment oil*, the marvellous *cough syrup*, the delicious *drops*. All these remedies guarantee an immediate and absolute cure for liver, kidney, and intestinal disorders, headaches, heart pains, stomach pains, sore throat, toothache, rheumatism, all sorts of pain, scabies, temper, ringworm, pox, gout, measles, as Molière said,[11] overwhelmed by the artistic-literary leaflets of pharmacists. In fact, nowhere is medicine more within everybody's reach than in those countries where the word '*the*'s pronounced with a lisp. What an enormous simplification, what a benefit in time! You could be out for a walk, at a dance, or going about your business and you feel out of sorts; you might be on foot, on horseback, or in a carriage and an accident happens; you may fall, break a bone; no matter what, the cure is ready; you reach into your pocket and swallow it, whatever one it is, since they are all good for whatever ails you, to get you back on your feet, as Molière would have said, laughing at popular English pharmacopoeia. After all, faith is a marvellous thing.

That reminds me of a story that happened to me when I was among the Chippewa Indians in the wilds of the Minnesota, near the headwaters of the Mississippi.[12] From the beginning of the expedition, the suffering that I and my companions had to endure was so bad that one of us, so disgusted by his apprenticeship to life in the wilderness, seized the opportunity that presented itself in the form of an Indian canoe that we met by chance as it went south.

He left, returned to St Paul, and died the next day. This created some alarm, so a box of medicine containing a few well-known remedies was prepared and sent to us. It was a complete collection of remedies that were as varied as they were unknown. The sight of all those mixtures, whose labels made too many promises to be sincere, filled us with respect and especially prudence; so we returned them all to the crate and did not concern ourselves with securing the lid. In the swamplands, our baggage and supplies were carried by porters, *half-breeds, Métis* Indians who differ from real Indians only inasmuch as they sometimes work. After a certain time, we thought to conduct an inventory of the crate: it was empty. The porters had eaten all of the pills and drunk all the potions, among which were a litre of tincture of arnica[13] and a bottle of "extract of Saturn."[14] Their health was none the worse. The investigation which followed showed that most of the above-mentioned remedies contained alcohol and were treated as fine liqueurs. Though we may be in good health today, of course, no one knows what the future will hold, and so it is appropriate to take precautions in advance.

In Sydney we see the officers of the *Minerve* and the *Talisman*. Having just arrived from the West Indies, they savour the coolness, which we find oppressively hot, having just arrived from the ice fields. We shake hands joyously; we had previously met two or three years ago in Japan, Madagascar, and Tahiti, and we meet again in Cape Breton. Who knows where the next meeting will occur?

The walks around the town are most enjoyable. I go down Main Street and turn right. I cross a small bridge over the entrance to a small cove where the houses and cottages that are half hidden by the fir trees are reflected in the water. Boats are tied up to the pilings, which support the banks. A line of ducks, quacking noisily, swim around until they feel tired, then rest in the sun. On the opposite side, we see Spanish River, which serves as a harbour, where French and other ships lie peacefully at anchor; beyond them on the other side, more cottages, farms in the middle of fields, surrounded by forests, and on the horizon a chain of low hills.

The water is beautiful, calm and clear, truly the mirror so dear to poets, as the expression goes, though it has been used and overused so much that we avoid it. The sky may not have the intense blue of the Mediterranean but is nonetheless pure and delicate and enhanced by the traces of clouds with their brilliant

Cape Breton. Street along Spanish River

Cape Breton. House on Spanish River, South Sydney

white edges, while their interiors are shaped of grey furls that turn almost reddish-brown in the centre. This splendid summer's day, this deep green vegetation has followed the melancholy; the cold and the fog of northern Newfoundland make up one of the most striking memories of the entire voyage. Beyond the point, the cottages are spaced farther apart; children play on the lawns – there are lots of children here – you can hear them shout, laugh, and diligently carry on with their occupations, sometimes to the sound of their mother's sewing machine, as she sits by the open window. On the sides of the road here and there, you see a lonely fir tree or a dried trunk, a testimony of the forest that formerly occupied this site and that civilization has left behind, to make use of its shade perhaps or out of scorn for those old, moss-covered patriarchs. On the other side, the telegraph poles are lined up one after the other, carrying a great many wires with their blue or green glass insulators. I telegraph from Sydney to France at nine o'clock in the morning, and the same day, by four o'clock in the afternoon, I receive a reply for only sixty *centimes* per word!

A short distance away, we meet the settlement of a Mic-Mac family; in the centre of a clearing stands the wigwam, about twenty poles joined together at the top to form a conical frame covered with strips of bark. However, the opening that serves as a door is barricaded; the family no longer lives in this hut, which must be a very poor shelter in winter; with pieces of board, a few sheets of metal, scraps that have been blown off houses in the neighbourhood, they have built a cabin modelled after an English house. That is the first step from aboriginal life to civilization. Such a sight would cause the philosopher to rejoice if the first step were not so likely to be the last. With very few exceptions, it requires such an effort on the part of an Indian that he dies. The men are absent or sleeping; we see only two women outside. An old woman crouches by the fire tending a pot held by three branches stuck in the ground; a younger one, with pronounced cheekbones, a reddish complexion, and black hair in braids falling over her shoulders, is dressed in a violet, iris-coloured calico skirt and a canary yellow camisole. While I observe her, a man comes by carrying a box containing an image that can be seen by looking through a lens over an opening; this is the very infancy of optics, optics for children. The Indian woman calls the man and after putting a new ten-cent piece into his hand, the price of the show, she puts her eye to the lens. I walk away, and as the road turns a corner, I notice her still looking at the same image.

I do not know what these Indians live on; they frequently move, since the forest is large, in spite of being cut little by little every day. The women make certain crafts of mediocre quality, moccasins embroidered with beads, baskets, cases for holding scissors made from porcupine bristles dyed different colours. These poor people, miserable fallen kings of the wild, have been transformed into bohemians who are as unable to follow the civilization that drags them on as they are to resist it; they die one after the other, proving by their deaths that inferior races of humans do not have the ability to change as quickly as would be needed to adapt to the superior races. Their existence is a function of a number of circumstances which, when eliminated, lead to their annihilation. The Australian aborigine is extinct; the Canaque and the Redskin are dying out; the Turk will soon be no more than a historic memory – *Sint ut sunt aut non sint*[15] – and they can only remain what they are. Men and entire races disappear and become the dust that I tread on as I continue to walk under the burning rays of the sun that is destined to cool some day and to die like everything in the universe. Grasshoppers fly in front of me and make a clicking sound by shaking their wings, along with the melody of the rustling of the birch tree and the murmur of the fir trees in the breeze. The road curves obediently to match the terrain and is bordered by a zigzagging fence around newly cleared land, where the stumps whiten in the elements until they fall to pieces and are ploughed under. Then the forest will have disappeared and made room for the fields of barley, oats, potatoes, like those which surround me, or for the pastures where fat cows with bells around their necks will graze and raise their heads and pink snouts to moo when they hear the sound of a buggy with fast wheels drawn by the trotting of a small, admirably muscular horse.

Another road leads out of Sydney, meets the crest of hills, and continues almost entirely through the woods to Louisbourg. About halfway there lies Miray, near a stream with no other current than the tides, widening in places to form lakes that enter the sea on the east coast of Cape Breton Island. You go over a bridge, past a church built of wood perched on a hilltop overlooking the lake; its pointed steeple, its white walls with five high-arched widows, are reflected in the water as it lazily laps onto the beach and

Cape Breton. Mi'kmaq Indian hut near Sydney

Cape Breton. Village of Miray [Mira]

Cape Breton. Spanish River near Fork Lake

against the side of a boat. The grassy mound serves as a pedestal, and the forest starts immediately behind it. This is no longer the Newfoundland forest, impenetrable and wild, where live and dead branches intertwine, where moss and brush block the light and air, and where moisture from the sky accumulates; this forest smacks of the proximity of civilized man. The ground is dry, the fir trees stand in uneven clumps, leaving room for open spaces around them, connected by lawns and paths carpeted by thick grass as short as the mountain grass in Cantal,[16] where herds of cows graze. This place is ideal for a picnic, the country pastime so well liked by the English. The members of various Sydney walking clubs miss no opportunity to avail of it; if the five officers of the *Clorinde* who spent such an enjoyable day there ever happen to read these lines, I am certain that they will recall it with pleasure. Miray [Mira] and Spanish Lake, near Forks River, are two favourite places. After the church, the road once again enters the forest, and three or four hundred metres farther you reach the village of Miray.

One must not try to compare an Anglo-American village to a French village. In the latter, thatched-roof houses, whether old or new, are built in a row and preferably close together and very often even touching one another, thatched roof against thatched roof, each house with its manure pile, its duck pond, and its garden fenced in by a hedge. It is not clean, but it is picturesque. The village has its streets, alleys, square, dead ends, and houses built around the church. The English farmer, on the contrary, hates straight lines and houses clustered together. He is unafraid of the open air and does not mind being a little isolated, provided he is a reasonable distance from his neighbours. As a result, his village is composed of houses that are so scattered that it is often difficult to see more than a half-dozen of them at one time. The spirit of a race manifests itself in the most minute ways. With only bone fragments, Cuvier[17] reconstructed in his mind an animal that had been extinct for thousands of years. By examining a grain of sand, a geologist can ascertain all the phenomena that it encountered and that left an imprint on it; as with the knowledge of a custom, the appearance of a work of art or of civilization indicates something about the character and the aspirations of the group of men who created the work. Is it not true that the history of Germany is all summed up in one of those heavy two-handed *lansquenet* swords? A piece of Della-Robbia[18] defines all of the Italian

Renaissance; the simplest jewel tells the entire history of Russia; and Miray is the quintessential Anglo-American village. The scattered houses, the two churches, make it English; the fences bordering the fields, the telegraph poles, the parallel lines of the clapboard walls, make it American; the woods around it make it North American. A single glance is enough to indicate the nationality and geographic location, which was what was to be shown, as the mathematician would say in concluding.

North Sydney is about three miles from South Sydney; you go there by taking a steamboat which stops only once, at the coal wharf. While my travelling companions and I enjoy the trip, which lasts three-quarters of an hour, we study the religious images, the pamphlets posted in the salon of the ferry; the accompanying text contains very respectable religious thoughts expressed in a more edifying than literary way; while we chat about our surroundings, today's walk and tomorrow's, three of our neighbours, very polite-looking gentlemen, watch us intently, and one of them finally asks me if it would be possible to visit the frigate. After our reply to the affirmative, he tears a sheet from his notebook, writes a few words, and asks us to accept two free tickets to the circus that is in town to give a great show later that day. Our three gentlemen, who are the clowns in the show, take leave of us at the wharf in North Sydney with friendly handshakes. How fashionable that is in the New World!

North Sydney is more American than English, South Sydney more English than American. The main coal mines are very close, and the population of miners with their boots, their sluggish way of walking, typical of people more used to using their arms than their legs, some with a small lamp attached to their hats to light their way in the mines, these engineers or foremen, sailors, businessmen who earn their living directly or indirectly from the coal mines, the women, the blond *girls*[19] with straw hats, Indian skirts with a floral pattern, and white aprons with bibs; everyone going about his business, walking, running, galloping on the board sidewalks, which reverberate beneath their heels, all bring North Sydney to life. In the streets, the hotels, boutiques, and houses all have flagpoles, large multicoloured signs, carriages, quite a sight for the eyes and an uproar for the ears. It is gay and lively.

After lunch – what a terrible lunch it was! I would gladly accept everything about England and America except the food – a buggy takes us to the mines, of which the largest is a few hundred

Cape Breton. View of South Sydney, from cemetery

metres from the sea and has a huge smokestack, various green-ish limestone buildings, and wheels turning so fast that the spokes become invisible. It would look quite mediocre next to the great crushers in the coal-mining regions of Pennsylvania. The coal, which is Cape Breton's great source of wealth, is a very brittle variety that gives off clouds of heavy black smoke when burned, contains a high percentage of pyrite, and is overall a very inferior-quality fuel. But because Sydney is the only deposit and the loading is easy, it has become a precious resource for the neighbouring regions of Newfoundland, Prince Edward Island, and Nova Scotia and the shores of the Gulf of St Lawrence when American coal, and especially New England coal, would be con-siderably more expensive. The deposit is approximately three feet thick, quite homogenous, with very little folds, and rising up in an even slope toward the east, where it is found on the surface. There are other, smaller deposits between Spanish River and Cow Bay on the Atlantic. The mine that we visit is totally dry, which makes it possible to dig tunnels under the seafloor to a depth of seven hundred feet.

When we are back in Sydney, we go to see the circus; for two hours we watch not so young performers, standing on the backs of horses, jump over banners and through hoops in time with the music while smiling to the audience and moving their arms

Cape Breton. Coal mine

Cape Breton. Coal wharf, North Sydney

in circles; then a man scratches his chin with the tip of his toe after first putting it behind his head; then he walks on his nose, makes braids with his legs, arms, stomach, and neck, and would make you stiff and sore just to watch him; another man, very well-dressed in red boots, a gold vest, and jacket, juggles very pleasantly with an infinite variety of objects, knives, bricks, balls; a clown in a red wig with his face powered white and the picture of the moon on one cheek and the sun on the other meows and receives a number of slaps and kicks and falls quite often, raising a lot of dust and creating an enthusiastic response from the younger members of the audience. The children applaud with frenzy, as we once did and as our great-grandchildren will do someday if, when they are no longer in diapers, they do not consider themselves too serious for such childishness and prefer to keep solely to politics.

The day of departure is drawing close; our hold is full of coal, the maple leaves are turning red, the squirrels have almost finished gathering food for the winter, and the English families will soon imitate them. Once again we hear the anchor chain wind around the capstan; the frigate sets sail, and soon the coast of Cape Breton Island disappears behind the horizon. We are en route to St-Pierre.

St-Pierre and the Trip Home

We find the island just as we had left it; the season has advanced, autumn has arrived, the temperature has cooled, and the fog now covers the mountain with its grey folds almost every day. I return to the places I have already visited and bid adieu to Cap à l'Aigle, the Iphigénie Monument, Savoyard, the *barachois*, and the Galantry lighthouse, still atop its rock pedestal with the surf beating against it. The sailors go ashore and take advantage of their leave to pay their respects to all the bars. As we walk the streets, we hear the melancholy strains of sentimental romantic songs drifting from an open window, sung by a powerful and quavering voice with a slow and monotonous rhythm, impossible for a real singer who might wish to amuse himself by imitating it, with its sudden crescendos and woeful endings, where love is inevitably betrayed. He sings of death, broken hearts, mountains, and tears. The sadder the song, the more beautiful. When it is time to return on board, the men come back up Gueydon Road, arm in arm, or with their hands in their pockets, staggering from side to side like a ship tacking against the wind. They all talk together without raising their voices but without stopping, confiding their sadness to the rocks, to the flakes where the fish are drying, giving lengthy explanations to the cod, struggling to keep their balance, repeating the same sentence over and over until a comrade comes to help them along a little farther. They finally reach the dock and get aboard the small boat that will take them back to the ship. A good night's sleep will put everything back in order, and tomorrow these big, friendly lads will go quietly back to their usual chores and count the days until their long-awaited, final liberation.

Before departing, we have invited a few guests to dine on board with us. For this solemn occasion, the entire mess and the chief steward have dressed in their most formal attire, and there are flowers in the vases; the décor is unanimously found to be magical, and we sit down at the table certain that we will do justice to the sacred duty of offering hospitality. The soup is mediocre. The steward then brings the roast duck, our main dish. The cook had examined the ducks that morning and had told us how marvellously plump they were. As soon as they are carried to the table, you can smell a hint of vanilla; when they are laid on the table, a scented cloud wafts around them, smelling worse – if it is possible to imagine – than a perfume shop. The person at the head of the table had watched the ducks being prepared for roasting, and had seen three pods of vanilla being removed from their insides. An investigation showed that the cook had gone ashore that afternoon. As he tells us this, we all look at one another in consternation. At this point we abandon all hope – the sad truth is obvious to us. The rest of the meal continues in the same vein: after duck *à la vanille*, we have caramel salad, mustard lentils with apple marmalade, rice cakes with lard and anchovies, and chocolate cream with Dutch cheese; it is an orgy of the most sundry culinary combinations, unheard-of dishes that mix garlic and sugar; spices, meat, fish, and vegetables have been blended and are astonished to find themselves joined together by the will of a drunken cook who has been seized by a cooking frenzy to mix, mix, and mix some more. Everything is mixed with everything. Once the first few moments pass, we reconcile ourselves – alas, there is no choice! The dishes are served one after the other; each one is tasted, usually provoking a series of grimaces as we try to determine what it contains; when it is discovered, bursts of laughter follow, and much discussion. Although our stomachs are not overtaxed, we digest the dinner thanks to a good measure of wit and humour, of which there is fortunately no shortage on board the *Clorinde*. The cook spends five well-deserved nights in chains to pay for his deed.

The cook is a curious person to study on board a ship; his authority and influence are far more considerable than you might be tempted to think. He has to be reckoned with, and there is not a midshipman, past or present, who can't recall an instance of a conflict in which a cook prevailed over a captain or even an admiral. Those rulers in white always have a ready-made vengeance at their fingertips; the fear of a sabotaged or even cancelled dinner on the occasion of an official invitation, the thought of a long voyage with deplorable dining, is enough to strike fear in the heart of any officer normally brave enough to remain calm in a storm or even face cannonballs.

The bad weather associated with the equinox makes its appearance; the fishing season is now over. The *Clorinde's* voyage is reaching an end, and we will soon start our long return journey to France. Everyone is delighted. One man thinks of those who await him; another thinks of those who will not be there to greet him; a third thinks of the European winter, which is milder than the one here in these icy regions; another is happy to return simply because he left, and he will be happy to leave again once he returns. Man has in him a taste for change; when it is today, he feels the desire, as powerful as an instinct, to dream of tomorrow, which so seldom keeps the promises it seemed to make and which takes him one step closer to death. In spite of my joy, I cannot help feeling moved. Every separation is painful, if not cruel; a land that I thought I was indifferent to, as I look at it now for the last time, sings a sad song of *adieu* to my eyes and heart. I realize as I go that I have left behind small bits of myself, like a poor sheep leaving a few flecks of his wool on the bushes he has passed. The crew is filled with joy, and the bugle never sounded as loud or as clear; the drum never rolled any better, and the sailors never turned the capstan as fast as they do the morning we raise the anchor to leave America and sail for Brest.

The sea is beautiful, the sky blue, and a strong southwest breeze carries us lightly at a speed of nine knots, with no need to turn on our lights. We are out of sight of land, and from the top deck, all that the eye can see are waves. We adjust to life on a long crossing, dinner following breakfast and followed by dinner; the officers start their regular watch. For the first few days, we find it difficult to adapt to the monotony, but soon we do; the clocks are adjusted, and we set our routine just as we put away our clothes. Although our thoughts are rocked by the ship's movement, they are still active. Life in a college, a barracks, or a monastery is only hard at the beginning; soon it becomes tolerable and even necessary. The body passively obeys the acquired habit while the mind, once it is released from minor occupations, is able to think freely.

   The weather deteriorates and we encounter rain shower after
rain shower; the wind strengthens, and the *Clorinde* lists severely,
rolling and pitching. The scuttles in the mess are closed, and the
only light is what comes in through the six brushed-glass lenses
that block the view and produce greenish semidarkness whenever
a wave breaks over the stern and plunges them under water. We
recline on the couches near one of the openings and read by the
weak light it provides. The confined air of the room where we eat,
drink, and even sleep – since several of our number have hung
their hammocks there where two large lamps are burning continuously – ends up being unfit to breathe. And since a heavy wave
has tipped over and smashed several bottles of liquor that were not
secured, the smell of chartreuse, cognac, and absinthe filters
through the crates and mixes with the noxious atmosphere. The
storm begins. On deck, the wind is blowing so violently that it is
dizzying, and covers your clothing with a cold spray. At first, we try
to eat at the table; each utensil, fork, spoon, knife, glass, plate,
serving dish, carafe is carefully held in place between pegs, but if
one of them suddenly breaks loose, it flies through the air like a
projectile, shattering glasses and dishes and bottles and spilling
the contents on our knees. We hold onto ropes that are strung
from the ceiling with wooden handles like those found over hos-
pital beds. To avoid tipping over, you have to wind your leg around
a leg of the table or hold on with both hands and wait in that position until the movement stops to drink or swallow a mouthful. It
is quite funny for a day or two but quickly loses its charm. When it
becomes impossible to sit at the table, we lie on the divans, bracing ourselves with cushions, and eat cold preserves. We do not
even try to carry out the slightest task. Seeing every object that is
laid down, even for a second, fly through the air, is a torture that
can only be fully appreciated when it has been experienced. Since
the beginning of the bad weather, every article in the cabin that
was not carefully stored has gone whirling madly around; the only
remedy is to stuff them every which way into drawers. As a result,
you do not know where anything is and you dare not touch anything for fear of disturbing the pile, which is the only thing guar-
anteeing stability. One more problem arises along with all the
others: the ship's hull has begun to leak, and along the walls, water
oozes and drips onto the furniture, the clothes, and the bed; you
try to fill the seam, but nothing stops it, so you go back to the mess
and try to read, sleep, and not lose your patience.

During the storm, the nights are as difficult as the days. The only rest you get is in a hammock. If you try to sleep in a bunk, every time the ship rolls, you are thrown against the side wall or, if you are not holding on, onto the floor. When sleep relaxes your grip, you are awakened by a brutal shock. With your eyes open and burning from insomnia, you make out the glow from the lamp drawing a line across the ceiling that oscillates continuously. If you close your eyes, your ear follows the groaning of the woodwork and the ship's rolling, followed by the crushing monotone that is impossible to grow accustomed to because you are still gripped by fear. The wind makes long, long cries like an enormous monster which roars tirelessly. When a huge wave rises from the dark depths and crashes against the side, next to your head, the entire ship vibrates, then is still for a moment and you tremble right to your bones because you are part of the ship, this living creature that is carrying you and protecting you from the storm. The hours pass ponderously. Your mind, your senses are dulled in that state between sleep and wakefulness; exhausted from the struggle, you are surrounded by horrible ghosts, terrifyingly real dangers, in a dark torpor, with no notion of time, space, past, present, or future, lost in the middle of a soft, colourless, shapeless cloud, looking but not seeing, listening without hearing, as though you were rolled up in huge, tangled circles, and · turning all together in the centre of an endless swirl which moves around silently until dawn.

The main preoccupation is whether the barometer is rising or falling, and whether the wind is going around to the east or to the west. If it is east, then we are being pushed away from France, west and we would only need a day or two to reach Brest, and then the wind can blow as hard as it wishes from any direction. The storm worsens. We are together in the mess, shattered from fatigue and the continual efforts needed to keep our balance; every minute, one of us gets up and clinging to the walls goes to the barometer hung in the corner, taps it with his finger to help the needle overcome inertia, and then studies the line drawn by the stylus on the paper, which moves too slowly on its roll. If he happens to trip and fall on his way back to the divan, no one laughs; he sits downs again with no other thought than to count the hours, the minutes, the seconds. One morning before four o'clock, a detonation comparable to a cannon being fired, followed by a shock, makes the entire *Clorinde* shudder. I immediately understand that something serious has just happened. All of a sudden, a deluge of water rushes in through the wall separating the gunroom from the deck, flooding the cabins, flowing from one side to the other with each roll of the ship. Everybody is up, wondering what is happening. We hear the order to secure the hatches; a wave has smashed the scuttle in the doctor's cabin, twisted the iron legs of the bed on which he was lying, and hurled him against the bulkhead, where he lies stunned by the avalanche that has swallowed him up. The wave have crashed through the opposite bulkhead and spread through and flooded the gun deck and the orlop deck. A whaling boat has been washed overboard, two other boats are damaged and are hanging eight or nine metres above the waterline, a yard has been broken, an anchor cathead damaged, the port gangway torn off, the captain's awning twisted, and six men injured.

The night is black, the wind howls, the sea roars. The carpenters are nailing heavy oak timbers behind the scuttle and bracing them against solid stays; the sound of their hammers continues evenly, without hurrying; there is no time for haste. Then a rope mat must be placed outside the scuttle to absorb the shock of the waves. This is a dangerous job. When a volunteer is called for, six come forward; the cabin boy is chosen and suspended on ropes outside the ship, in the most vulnerable place of all because it has just been struck, with more than a risk of being crushed by the mountainous waves so high that in order to see their tops from the poop deck, we must raise our heads. In the middle of this howling hurricane, so terrible that a shout cannot be heard more than a few metres away, he attaches the ropes that hold the mat; when the job is done and he is untied, he goes back to his post. When you think about it, there are clever people who find a way to earn one hundred thousand pounds in income, without counting the positions and the honours, with a *sou*'s worth of dedication skilfully invested, while honest sailors, valiant, noble imbeciles, sacrifice their health, their limbs, their lives, more than one hundred thousand francs worth of dedication, all of that for one *sou* a day and food. What good, brave people! The better you know them, the more you honour and love them!

The storm lasts two days longer; for seven days now we have been holding the same course. The wind drops slightly but is still so strong that the waves are exactly the same height; whenever one of them rises above the others that protect it, its crest is immediately shaved off; we resume our course in heavy seas. The barometer rises, the fine weather returns. We steam full speed ahead for fear of encountering another gale. The old *Clorinde*, like a horse that smells its stable, pushes its wide cheeks through the swell. Our eyes scan the horizon anxiously to spy a dark cloud that appears, but especially straight ahead looking for land. A white dot is spotted; it is the Ouessant lighthouse! We get closer, night falls; on the coastline, lights appear along our side; we pass Molène Island, Pierres-Noires, Toulinguet Point,[1] the Chaussée de Sein; we cross the narrows and enter the harbour; the town and its lights look like a sparkling amphitheatre. "Port side, drop anchor! Starboard, drop anchor!" orders the captain. The chains unwind, the anchors drop, we have arrived.

And now, after the last night aboard, I go down the ladder, board the rowboat, and leave the *Clorinde*, each stroke of the oars taking me farther from her; I look at her now, motionless, her tall mast standing straight up in the middle of the harbour. My heart is heavy as I leave the ship that has carried me for six months through wind, fog, sunshine, rain, calm, and storm. I shake hands with the officers, my dear companions on this journey. May they feel the same affection that they always showed toward me, because I found them always so benevolent and kind. They paid me the honour of welcoming me as though I were one of them, and for that I am deeply grateful. Together, we have been devoured by mosquitoes, together we have been cold and hot. Around the same table, we have shared meals both good and bad; in the same mess, we have exchanged ideas, talked of our hopes and occasionally of our worries. Each of us will now go his own way. For sailors, an arrival is always followed by another departure. As soon as they are discharged from this vessel, they will scatter to the seven seas. Wherever they may go, I hope that these lines come before their eyes and testify to the fond memories I have of them; may my words give them as much happiness as they do to me as I write them and remind them of our cordial relations all throughout the *Clorinde's* Newfoundland voyage during the summer of 1886.

Appendix
Official Correspondence

1 Nancy, 1 March 1886. Thoulet to the Director of Higher Education requesting leave and permission to travel to Newfoundland to conduct research.

2 Nancy, 3 March 1886. Grandeau, Dean of Science at the University of Nancy, to the Superintendent of Education, District of Nancy, recommending that Thoulet's request be granted.

3 Nancy, 4 March 1886. Maurice, Superintendent of Education, District of Nancy, to the Director of Higher Education recommending that Thoulet's request be granted.

4 Paris, 13 March 1886. Minister of Education to the Minister for the Navy and the Colonies requesting permission for Thoulet to embark on a vessel of the French Navy.

5 Paris, 18 March 1886. Minister for the Navy and the Colonies to the Minister of Education, granting permission for Thoulet to travel on the *Clorinde*, costs to be borne by the Ministry of Education.

6 Paris, 22 March 1886. From the Minister of Education to the Superintendent of Education, District of Nancy, informing him that Thoulet has been granted permission to travel aboard the *Clorinde* to depart on 1 May, with dining privileges at the officers' table.

7 Nancy, 23 March 1886. Thoulet to the Director of Higher Education, raising the problem of his not being able to afford to pay for meals on board.

8 Nancy, 23 March 1886. Dean of Science to Director of Higher Education, asking him to intervene on Thoulet's behalf and request that the Minister waive the cost of meals.

9 Paris, 26 March 1886. Minister of Education to the Minister for the Navy and the Colonies, who requests that, given Thoulet's lack of funding and the importance of his work for the Navy, his colleague re-examine his earlier decision and waive Thoulet's expenses.

10 Paris, 31 March 1886. Minister for the Navy to the Minister of Education. Under the special circumstance, meals will be provided for Thoulet and the cost paid by the Navy.

11 Paris, 9 April 1886. Minister of Education to the Superintendent of Education, conveying the favourable decision that the Minister for the Navy will assume the expenses for Thoulet's meals aboard the *Clorinde*. Thoulet must be present in Lorient on 29 April.

LETTER 1
Thoulet to the Director of Higher Education

Nancy, March 1st 1886
Sir:
You are, of course, aware of all the attention that has been paid recently to the study of the world's oceans, whose importance most governments understand full well. The expeditions carried out by the *Travailleur* and the *Talisman* have followed the example set by the Americans, the English and the Swedes. However, in France, scientists have limited their work to the field of marine zoology and, by and large, physics and chemistry have not been given the systematic treatment they require. The problems that need solving are numerous and of primary importance to applied sciences like meteorology and hydrography as well as to pure sciences like zoology and geology. I refer only to those problems concerning oceans currents, the composition of sea-water at different depths, the ratio of the salt and carbonic acid it contains, the solid particles suspended in it, and finally, the detailed analysis of the mud and sand brought up from the sea floor. Geology has as one of its principal aims to study the sediment deposited during the early eras of the earth's formation. It is surprising that, in the case of phenomena which are occurring in oceans even today, so little is known.

For a long time now, I have been involved with experimental geology; the results that I have obtained can be found in a number of published works. I now recognize the necessity to seek further knowledge through field work that can only be carried out on board ship, in entirely different conditions from those in our laboratories. With this objective in mind, I am requesting your support as well as that of the Minister of Education and the Minister for the Navy, in obtaining authorization to embark upon a voyage aboard one of the Navy's vessels stationed along the coast of Newfoundland. This region presents a particular interest precisely from the point of view of the questions which my own research has made me most familiar with: the study of the sediments that have been carried by the polar ice-floes and deposited to form the Grand Banks of Newfoundland, as well as the erosion, caused by frost, of the rocks along the northernmost part of the coast of the island.

In the event that the Minister for the Navy agrees to grant me permission to travel aboard a vessel of the navy and to dine at the officers' table during my trip, and that the Minister of Education grants me leave of absence for six months without loss of salary, in short, if this expedition does not require any additional expense above the limit that my financial situation would allow, I would be happy to carry out such a mission. Upon my return, I could then proceed to undertake a detailed study of the samples that I would bring back.

Most respectfully yours,
J. Thoulet, Professor of Geology & Mineralogy
Faculty of Science, Nancy

LETTER 2
The Dean of Science, University of Nancy, to the Superintendent of Education, (*recteur*), District of Nancy

Nancy, March 3rd 1886
Sir,

I am honoured to forward to you the attached application which my colleague Julien Thoulet addressed to the Director of Higher Education, requesting leave for a scientific expedition to the coast of Newfoundland where he would study marine (ocean) sediments.

Mr Thoulet has already discussed his wish with Mr Livet. My letter is to indicate to you that Mr Thoulet's absence would not cause any problem for the students who are preparing their degrees. Mr Thoulet has completed teaching their prescribed courses. The research that my colleague has already published relating to Earth Science is a guarantee that the expedition which he is planning holds much promise for further progress in this field.

L. Grandeau, Dean of Science

LETTER 3

The Superintendent of Education, District of Nancy, to the Director of Higher Education

Nancy, March 4th 1886

Sir,

I am honoured to forward to you the application from Mr Thoulet that you and I have already discussed.

I attach the letter of approval from Mr Grandeau, Dean of Science.

I myself would be delighted if Mr Thoulet's request could be granted. He is a distinguished professor who is already well known for valuable work. In my opinion, he is in every way deserving of favourable consideration from your administration. In requesting the means to carry out such a long and difficult voyage, his only ambition is to further scientific knowledge, to which he has devoted his life.

M. Maurice,
Superintendant of Education, District of Nancy

LETTER 4

The Minister of Education to the Minister for the Navy and the Colonies

Paris, March 13th 1886

Mr Minister and Esteemed Colleague,

Mr Thoulet, Professor of Geology and Mineralogy in the Faculty of Sciences at the University of Nancy has been conducting very interesting research in experimental Geology for quite some time and has published his results in various books that have been well received.

In order to complete his current work, Thoulet would need to carry out experiments that cannot be performed in the laboratories at his disposal and he would need to explore northern Newfoundland, specifically to study the sediments which compose the Grand Banks of Newfoundland.

In view of the foregoing, I am honoured to request your authorisation that Mr Thoulet be granted permission to embark aboard one of the vessels of the French Navy and to dine at the officers' table during his journey. Since this is a matter of great scientific interest to France, may I therefore prevail upon you to take a favourable view of Mr Thoulet's application?

I would greatly appreciate your informing me of your decision regarding this request.

I remain etc.,

The Minister of Education

LETTER 5
The Minister for the Navy and the Colonies to the Minister of Education

Paris, March 18th 1886
Mr Minister and Esteemed Colleague,

In reply to your letter of March 13th, I have the honour of informing you that I have instructed the *Préfet maritime (3e arrondissement)* to arrange for Mr Thoulet, Professor of Geology and Mineralogy in the Faculty of Sciences at the University of Nancy, to embark aboard the *Clorinde*, with dining privileges at the officers' table, in order to continue his geological research.

As a public servant, the cost of meals on board for Mr Thoulet will be borne by the Department of Education, who will reimburse the Department for the Navy and the Colonies in the customary way.

Mr Thoulet is expected to present himself at the port of Lorient on April 25th at the latest.

I remain, etc.
The Minister of Education

LETTER 6
The Minister of Education to the Superintendent of Education, District of Nancy

Paris, March 22nd 1886
Sir,

I am honoured to inform you that, on my recommendation, the Minister for the Navy and the Colonies, authorizes Mr Thoulet, Professor of Geology and Mineralogy in the Faculty of

Science at the University of Nancy, to embark aboard the *Clorinde*, with dining privileges at the officers' table, to continue his geological research.

However, Mr Thoulet himself must bear the cost of his meals on board and he must be in the port of Lorient no later than April 29th.

I would ask you to please inform Mr Thoulet of this decision and impress upon him the urgency of the matter and the conditions under which he may embark.

I remain, etc.

LETTER 7
Thoulet to the Director Higher Education

Nancy, March 23rd 1886
Sir,

Today, I received notification that the request I had the honour of making to you to embark on board the French naval vessel the *Clorinde*, has been granted.

Unfortunately, I also learned that I myself must bear the cost of meals during my stay on board. I would like to draw your attention to the fact that, since I have not received any financial support, that even if my meals were provided free of charge to me, the cost of this expedition is becoming increasingly heavy. I would have to purchase a considerable amount of scientific equipment which is expensive and costly to transport. Moreover, my research is intended to study ocean currents, sea water and sediments, all of which is hydrographical in nature and of particular interest to the Navy.

I would be deeply grateful if you would entertain my request for favourable consideration and allow me to prevail upon you to request that you appeal to the Minister for the Navy to provide my meals free of change. If not, to my profound regret, I would be forced to decline this expedition which would be too onerous for me, but which, I cannot help thinking, would be of some usefulness to science.

I remain, etc.
J.Thoulet

LETTER 8
The Dean of Science to the Director of Higher Education

Nancy, March 23rd 1886
Sir,

Mr Thoulet, with the energy he applies to his scientific research, was preparing to spend six months in Newfoundland, <u>without funding</u>; he has just today received notification that his meals during the voyage would be at his own expense. He has written you the attached letter asking you to intervene on his behalf to the Minister for the Navy and the Colonies so that his food on board be free of charge. I would like to add my own earnest entreaties to his, because I am sure that his expedition will result in progress in this very important but little-known study of marine lithology. I would be grateful if you agreed to petition the Minister of the Navy, in the interest of science.

I remain, as ever ...
L.Grandeau

LETTER 9
The Minister of Education to the Minister for the Navy and the Colonies

Paris, March 26th 1886
Mr Minister and Esteemed Colleague,

I have informed Mr Thoulet of your favourable decision to authorize him to embark on board to Clorinde, enabling him to explore the northern part of the island of Newfoundland, on the condition, however, that the cost for his meals on board be borne by my budget.

I had no choice but to inform Mr Thoulet that it was not possible to fund this expense from my budget. Mr Thoulet wrote me today stating that the expedition in question, for which he is receiving no remuneration, would require him to purchase a considerable amount of scientific equipment which would be very costly to transport and that, combined with the cost of meals would far exceed his resources.

Under these conditions, I would like to ask you to re-examine the request that I had the honour of submitting to you on behalf of Mr Thoulet and I appeal to you to grant that his meals be provided to him free of change. I would add that Mr Thoulet's expedition will concentrate on the important scientific question of ocean currents which is of particular interest to the Navy.

[Signature]

LETTER 10
The Minister for the Navy to the Minister of Education

Paris, March 31st, 1886
Mr Minister and Esteemed Colleague,

You informed me in your letter of March 26th, that it would be impossible for your Department to cover the cost of meals for Mr Thoulet during his stay on board the *Clorinde.* You pointed out also that he is not receiving any additional remuneration for this expedition and has had to purchase expensive scientific equipment, the cost of which, added to that of his food, exceeds his resources.

Mr Thoulet's expedition, as you yourself point out in your letter, will study an important scientific question. I have therefore decided that, in this exceptional case, and contrary to what I stated in my letter of March 18th, the cost of Mr Thoulet's meals on board the *Clorinde* will be borne by the Department for the Navy. I would be grateful if you informed Mr Thoulet of this arrangement.

I remain, etc:

LETTER 11
The Minister of Education to the Superintendent of Education, District of Nancy.

Paris, April 9th, 1886
Sir,

Following my letter of March 22nd, I am honoured to inform you that, at my insistence, and in this exceptional case, the Minister for the Navy and the Colonies has decided in favour of waiving the cost of meals for Mr Thoulet during his stay on board the

Clorinde where he has been given authorization to embark in order to explore the island of Newfoundland; the cost will be borne by the Department for the Navy.

I would therefore ask you to inform Mr Thoulet of this generous decision and remind him of how advantageous it is. He must present himself at the port of Lorient no later than April 29th.

I remain ... etc.

Tables

TABLE 1

Serial publication data for *Un voyage à Terre-Neuve.*

Source: *Bulletin de la Société de géographie de l'Est.*

Chapter Headings	*Bulletin de la Société de géographie de l'Est* (6 installments)	Berger-Levrault, 1st ed. 1891
Lorient	Vol. 12 1890, 1–8	1–8
The Crossing	9–24	9–24
The Islands of St.-Pierre and Miquelon	24–9 vol. 12 1890, 236–48	24–42
A Little Geography and History	248–64	42–58
Bonne Bay	vol. 12 1890, 459–72	58–71
Ingornachoix Bay, Port Saunders and St Margaret Bay	472–83	71–82
St Lunaire, Croque Harbour, and the Mosquitoes	vol. 13 1891, 1–17	82–99
Jacques Cartier Bay and Sacred Bay	17–31	99–113
The Cod Fishery	vol. 13 1891, 487–503	113–29
The Eastern Shore and Cat Arm	503–14	129–40
Labrador and Cape Breton	vol. 14 1892, 26–47	140–61
St Pierre and the Trip Home	47–56	162–71

TABLE 2

Photographs published.

Source: J. Carpine-Lancre

Thoulet's photos	Un voyage à Terre-Neuve Paris; Nancy: Berger-Levrault, 1891	*Bulletin de la Société de géographie de l'Est*
St-Pierre. Wharves near cap à l'Aigle (not included in Thoulet's Album)	Frontispiece	Vol. 12 1890 between 474 and 475
St-Pierre. 07. La Roncière Quay	Between 30 and 31	Vol. 12 1890 between 24 and 25
St-Pierre. 09. Cod drying on rocks and on "bordelaises"		Vol. 12 1890 between 24 and 25
Newfoundland. 35. Shoal Brook Falls Bonne Baie		Vol. 12 1890 between 262 and 263
Newfoundland. 37. Shoal Brook Bridge Bonne Baie		Vol. 12 1890 between 474 and 475
Newfoundland. 34. Eroded rock. Ingornachoix Bay	Between 76 and 77	Vol. 12 1890 between 262 and 263
Cape Breton. 16. Main Steet South Sydney	Between 144 and 145	

Notes

INTRODUCTION

Except where otherwise indicated, translations are mine.

1 Carpine-Lancre, 2003b; letter to the author, 9 April 2000; and
  Carpine, 2002.
2 Thoulet, «Sept mois,» 601. See also Carpine-Lancre, 2003b.
3 *Bulletin de la Société de géographie,* vol. 13, 610–11 (1867). Gnomonic. From gnomon. "*Geom.* 2. The figure that remains after a parallelogram has been removed from the corner of a larger but similar
  parallelogram 3. *Astrom.* A column etc. used in observing the sun's
  meridian altitude" *(Oxford English Dictionary).*
4 *Contributions à l'étude des propriétés physiques et chimiques des minéraux
  microscopiques.* See also *Annales de physique et chimie,* 1880, 5e série,
  vol. 20.
5 Carpine-Lancre, "Julien Thoulet," 1.
6 Thoulet had reached seventy, so retirement was mandatory, in spite
  of his good health and his desire to continue working. (Letter from
  Thoulet to Albert I of Monaco, 27 November 1913, Archives du
  musée océanographique de Monaco.)
7 The term "océanographie" in French dated from 1584 but is considered "rare before 1876" by *Le Robert,* which defines it as "description et établissement de la cartographie des océans et de leurs fond
  marins." According to Carpine (*La pratique,* 160n3), "it is to a large
  extent under [Thoulet's] influence that the term 'océanographie'
  became popular in French." In English: "oceanography, that
  branch of physical geography which treats of the ocean. Its form,
  physical features and phenomena, date from the commencement
  of the *Challenger* investigations" (*OED,* vol. VII, p. 50).
8 Thoulet, 1884a, 1885a, 1885b.
9 For this biographical sketch of Thoulet, I have relied heavily on information supplied by Jacqueline Carpine–Lancre, 2003a, 2003b.

10 In 1897 and 1898, Thoulet published his analyses of the sediments
he had collected in the Gulf of Gascony aboard the *Caudan* in
1895. See also Thoulet 1889a and 1889b.

11 Thoulet, *Échantillons d'eaux.*

12 Margaret Deacon, *Scientists and the Sea,* 392.

13 See Thoulet, 1901b.

14 Carpine-Lancre, "Origins," 14.

15 General Bathymetric Chart of the Oceans,
www.ngdc.noaa.gov/mgg/gebco/gebco.html.

16 Montauban: Association française pour l'avancement des sciences,
31e session, 2e partie, 1074–93, figs. 1 to 4. 1902.

17 When the context permits easy understanding of official administrative or political titles or terms, they have not been translated.

18 *Conseil supérieur des pêches maritimes.* Archives nationales, Paris,
F/17/22246.

19 Thoulet. Letter to Dr Jules Richard on 21 December 1918
(Archives Musée océanographique de Monaco). Thoulet wrote that
his wife and four children had all lived through the war and were in
good health. His younger son was an architect, and his younger
daughter worked as draughtsperson in a factory.

20 Letter from Thoulet to Sauerwein, 21 January 1903. Thoulet states
that he was told in Nancy that the problem lay in Paris, and in Paris
that the problem lay in Nancy. Bichat advised him to elicit support
in the Ministry of Education. Thoulet wrote to Sauerwein to request the help of his Highness, Prince Albert of Monaco, and did
so again nearly two years later. At the time, the Freemasons (to
which Sauerwein belonged) were using thousands of secret files to
investigate the religious beliefs and practices of anyone teaching in
France, for the purpose of promoting only those who were "un-
equivocally faithful to republican institutions." See Birnbaum,
La France imaginée, 174. Possibly, Thoulet was a victim of that anti-
clerical intolerance.

21 Carpine-Lancre, "Les croisières océanographiques" and "Les
expéditions océanographiques."

22 Known since the twentieth century as the Muséum national
d'histoire naturelle. www.mnhn.fr

23 Vessels such as the *Charente, François-Arago,* and *Pouyer-Quertier* plied
the Atlantic, South Pacific, and Indian oceans. Carpine-Lancre,
"Les croisières océanographiques."

24 For example, Louis-Antoine de Bougainville, aboard *La Boudeuse*
(1767–71); and Jean-François de Gaulaup, comte de la Pérouse,
aboard two ships, *Le Boussole* and *l'Astrolabe* (1785–89).

25 Carpine-Lancre, "Les croisières océanographiques," 6–7. See also
Folin, *Sous les mers;* and Filhol, *La vie au fond des mers.*

26 Carpine-Lancre "Les croisières océanographiques," 7–8. The field
of aeronautics would of course soon – after the turn of the century
– generate renewed confidence. See also below, 172–3n10.

27 Thomson, ed., *Report.*

28 Some years later, Thoulet published a detailed review of the scientific results of the Voyage of the *H.M.S. Challenger,* during the years 1873–76, by John Murray and Renard. See Thoulet, "Les dépôts sous-marins" and "Les derniers volumes des reports du *Challenger.*"

29 Thoulet, "Experiences relatives à la vitesse," 1513.

30 See Appendix, letter 7, in which Thoulet indicates in no uncertain terms that his financial situation would scarcely allow him to pay for a fraction of the cost of his Newfoundland trip.

31 It has since changed somewhat, but in Thoulet's time, the French education system had three parts: *enseignement primaire* (primary), *enseignement secondaire* (high or secondary), and *enseignement supérieur* (university or postsecondary). The President of the Republic, on the recommendation of the Minister of Public Instruction (Education), appointed the *recteur* (that is, superintendent) of each *académie* – that is to say, geographical zone (not a learned body or society).

32 See Appendix, Letter 1, 1 March 1886. Archives nationales, Paris, F/17/22246.

33 From 1783 to 1904, the West Coast and Northern Peninsula from Cape Ray (near Port aux Basques) to Cape St John (White Bay). See page 2 above.

34 Carpine, *La pratique,* 159–76.

35 For recent scholarly explanations of the origins and early history of the French Shore, see Hiller, "Utrecht Revisited," 23–39, and "The Newfoundland Fisheries Issue," 1–23; and Brière, "Pêche et politique," 168–87.

36 Appendix, Letters 1–10. Archives nationales, F/17/22246.

37 Appendix, Letter 6. 22 March 1886.

38 Appendix, Letter 7. 23 March 1886.

39 Appendix, Letter 9. 26 March 1886.

40 Appendix, Letter 10. 9 April 1886.

41 See below, page 10.

42 Lefort and Lemesle, *La Royale et les Terre-Neuvas,* 128.

43 See page 11.

44 Lefort and Lemesle, *La Royale et les Terre-Neuvas,* 128ff.

45 1838–96. *Commandeur de la Légion d'honneur, Officier d'Académie* (1878), *Officier de l'instruction publique.* In 1861, aboard the *Pomone,* as Aspirant *1ere classe* [midshipman, first class], described as a "good cartographer"; in 1878, as *Second,* aboard the *Laplace,* collaborated on several maps of Newfoundland; in 1881, as *Commandant* of the *Indre,* was "commended for having rescued the *Ashburn* that went aground in Mall Bay in Northern Newfoundland." Promoted to the rank of *Captaine de vaisseau* in October 1881. French Naval Attaché in London, 1883–86. Appointed *Commandant* of the 6-cannon, 1200-horsepower warship *Indomptable,* 20 January 1887. (See Salkin-Laparra, *Marins et diplomates.*)

46 *Annuaire de la Marine et des Colonies, 1886.*
47 Thoulet, *Voyage,* 11: "Every summer for the last few years." According to Querré, *La Grande Aventure,* 197, the *Clorinde* started coming to Newfoundland in 1880 and continued until 1887.
48 Inscriptions on the rocks near Le Fond, Northeast Brook, Croque Harbour, show the names of several ships – the *Crocodile,* the *Ibis,* the *Pomone 1868,* and the *Roland 1862* – and the names of some crew members.
49 Koenig, "Le French Shore."
50 "Lettre du 28 octobre 1886," *Nouvelles de la flotte,* BB/2/640.
51 See below, pages 149 to 152.
52 C. Querré, *La Grande Aventure,* 197.
53 On page 6, Thoulet describes the market he saw on Easter Sunday in Lorient, which fell on 25 April 1886. Clearly, then, he had arrived at least a week before the scheduled departure. See also "Chapter 2. The Crossing," and Koenig, "Le 'French Shore,'" 370, which gives 1 May as the departure date. Although the ship's sailing was delayed for a few days, Thoulet begins his narrative as soon as he boards.
54 "Observations faites à Terre-Neuve," 398–430.
55 Thoulet (1888a).
56 Ibid., 314–23.
57 Thoulet had a particular interest in the colour of the ocean, which is a complex and fascinating question. Thoulet examined seawater colour in a number of his works, and in order to gauge and interpret colour, developed several devices, most of which used coloured gelatine sheets. These devices included a *tube colorimétrique* – a colorimetric tube; a *lunette colorimétrique* – a colorimetric viewer; a *chapelet colorimétrique* – a string of coloured squares; and finally a *planchette colorimétrique,* or colorimetric strip. For illustrations and explanations, see C. Caprine, *La Pratique,* 170–2; and *Catalogue des appareils d'océanographie,* 50–3.
58 There is no mention of the *Clorinde* (launched in 1845) in Bourges, *Journaux de bord.*
59 See page 99.
60 Koenig (1889), 385.
61 See pages 115 and 116.
62 I am grateful to Jacqueline Carpine-Lancre for providing detailed information regarding the publication of *Un voyage à Terre-Neuve* in serial form.
63 The final instalment actually appeared in *Bulletin de la Société de géographie de l'Est* in 1892, the year after the publication of the book.
64 Three of the photos had been published in the *Bulletin de la Société de géographie de l'Est* – "Le quai de La Roncière. St. Pierre," "Roche érodée à la baie d'Ingornachoix. Terre-Neuve," and "Main Street à South Sydney, Cap Breton" – in addition to another, the

frontispiece ("Les Cales près du cap à l'aigle"), not included in album described below.

65 Thoulet, "Sur le mode de formation des bancs de Terre-Neuve," 1042.

66 Thoulet, "Sur le mode d'érosion des roches," 1193.

67 Fiero, *Inventaire des photographies.*

68 Société de géographie, 184 boulevard saint-Germain, 75006 Paris. Thoulet's photographs are housed in the Bibliothèque nationale de France, Département des cartes et plans, rue de Richelieu, 75084 Paris Cedex 02.

69 See page 122.

70 Thoulet (1887a), 390.

71 *Bulletin trimestriel,* 316.

72 See page 39.

73 See page 74.

74 'New Land' – an archipelago in the Russian Arctic.

75 See page 126.

76 Charles Baudelaire, from *Les fleurs du mal* (1857): "Les parfums, les couleurs et les sons se répondent."

77 Arthur Rimaud, *Poésies* (1871).

78 See page 40. "All these islands and peninsulas … evidence of extensive sinking of this part of Newfoundland." On fjords, see also pages 54 and 55 and 118 to 119.

79 Appendix. Letter 1. Thoulet to the Director of Higher Education.

80 Thoulet (1889b) credits Prestwick with this theory.

81 Ibid., 222.

82 See page 54.

83 Perret, *La Géographie de Terre-Neuve.*

84 Ibid., 78.

85 See page 103.

86 See page 46.

87 For a summary of the history of the French presence in Newfoundland, see "France," "French Shore," and related entries in the *Encyclopedia of Newfoundland and Labrador,* St John's, 1981–94.

88 See pages 66 and 67.

89 After more than a century, Loti's book is still in print (it was republished in 1993 by Christian Pirot). It has been adapted for the screen at least three times: in 1924 by J. Baroncelli, in 1958 by Pierre Schoendoerffer, and in 1995 by Daniel Vigne. Available online at Gallica. gallica.bnf.fr/.

90 M.F. Maury, *Steam-lanes,* argues for a solution to a problem that has long been a topic of discussion at conferences dealing with navigation and rescue. See page 113.

91 See page 59.

92 See page 41.

93 Thoulet's family connections included the three brothers of his mother (Marie Pauline née Nisard): Désiré Nisard, a literary critic,

director of the Normal School, and member of the illustrious Académie française; Charles Nisard, a writer; and Auguste Nisard, a humanist, Superintendant of Education for the district of Grenoble, and Dean of the Faculty of Letters of the Catholic University of Paris. Having spent part of his childhood in Algiers, Thoulet probably did not meet or get to know his famous uncles until he came to Paris to complete his baccalaureate. However, it seems safe to assume that his mother benefited from the same classical upbringing as her brothers and would have seen to it that her sons received a well-rounded education. It is more difficult to ascertain the level of education of Thoulet's father; all we know is that he was in business and that Thoulet's grandfather was a man of independent means.

94 Manuscript letter from Thoulet to the Minister [of Education] of 29 January 1874. Archives, Collège de France, Paris, no. 2103.

95 Letter of 27 June 1891, Archives, Musée océanographique de Monaco.

96 Two articles in *Engineering and Mining Journal of New York*: "History of the Blowpipe," 9, no. 16 (1869): 243; and "Applications of the Pentagonal Symmetry of the Terrestrial Globe," 9, no. 9 (1879): 131.

97 *Le Gulf-Stream ... Bulletin de la société de géographie de Marseille*, 269.

98 See page 116.

99 It is possible that King Alphonse XIII of Spain used the phrase to describe Thoulet, who, accompanying Prince Albert of Monaco, took part in an International Commission of Scientific discovery held in Madrid in November 1919 under the auspices of the king.

100 *Journal de Monaco*, 46e année, no. 2323 (13 January 1903), 2. The following month, in an article in *Le Figaro* (49e année, 3e série, no. 33, 2 February 1903, 1–2), "The Future of Our Fisheries: The Opinion of Prince Albert of Monaco," the Prince described Thoulet as "the master who studies the waters of the deep and the ocean floor."

CHAPTER ONE

1 Victor Massé (Lorient 1822–Paris 1884). Composer of such popular comic operas as *Les noces de Jeanette, La reine Topaze,* and *Carnaval de Venise.*

2 Reddish flowers also known as "wild artichokes," commonly found on old stone walls. Latin: *jovis barba.*

3 "If it be allowable to compare small things with great" – Virgil

CHAPTER TWO

1 Between Corsica and Sardinia.

2 Whale.

3 Also spelled rep. "A silk, cotton, rayon, or wool fabric, having a cross-wise rib" – *Funk & Wagnall's Standard College Dictionary.*

4 French botanist (1801–32) and traveller in North America, India, and Tibet; author of *Correspondance.*

5 Or *wili.* Nymph, in Slavic mythology, "chiefly used in connection with the ballet *Gisele ou les Willis* (1841) by Vernoy de Saint-Georges" – *Oxford English Dictionary,* vol. XX, 340. Norn. "One of the female Fates recognized in Scandinavian mythology, chiefly in the plural." "These virgins are they who dispense the ages of men, they are called Nornies, that is fairies or Destinies." (Percy 1770. Tr. Mallet's *Northern Antiquity,* II, 51.) *Oxford English Dictionary,* vol. X, 518.

6 "In Norse mythology, each of Odin's twelve handmaidens who selected heroes destined to be slain" – *OED.*

7 Faun(?) "faune – god of fertility, of flocks, and fields," *Grand Dictionnaire Larousse de la langue française.* "In Roman mythology, a woodland deity typical represented by a man having the ears, horns, tail and hind legs of a goat, satyr" – *Funk & Wagnall's.*

8 "Dog Island" is now called Île aux marins.

CHAPTER THREE

1 For a comprehensive essay in French on St-Pierre and Miquelon, see Guyotjeannin, *Saint-Pierre et Miquelon.* In English (although the account is not as detailed), see Andrieux, *St. Pierre and Miquelon.* See also Lebailly's illustrated *Saint-Pierre et Miquelon Histoire,* and www.st-pierre-et-miquelon.com/.

2 North-East Channel, Halibut Channel, South-East Channel.

3 Victory, Pigeon, and Massacre Islands.

4 Green Island.

5 Workers, usually young boys who spread the cod to dry on *graves* (also *grèves)*—rocky beaches on which cod was dried. Brasseur and Chauveau, *Dictionnaire,* 377.

6 As ballast.

7 Greek god of Law.

8 Fontaine (1678), "L'Huître et les plaideurs," Fable IX, Livre VIII :

The Oyster and the Litigants
[Translated by James Michie (1979) *Selected Fables,* Penguin, 1982, 119–20]
One day two travelers, walking side by side,
Come on an oyster washed up by the tide.
Greedily they devoured it with their eyes,
Excitedly they pointed out its size,
And then, inevitably, they faced
The problem : which of them as judge
Should pass a verdict on the taste?
One was already stooping for the prize

When his friend gave him a nudge :
   'We must decide this properly.
   The epicure's monopoly
Belongs to whoever saw it first : *he* swallows
The oyster and, it logically follows,
   The other has to watch him do it.'
   'If that's the way you view it,
I have, thank God, remarkably keen sight.'
   'Mine's pretty good as well,
   And upon my life, I swear
I saw it before you!' 'So what? All right,
   You may have been
The first to *see,* but I was the first to *smell.'*
Who should arrive upon this charming scene
But Perrin Dandin? Asked to intervene
   As arbiter in the affair,
   With a portentous air,
He digs the oyster from its shell
And gulps it while his audience stands and stares.
   The meal finished, he declares
In the tone of voice beloved of presidents :
   'The court hereby decrees
An award, without costs, of one shell to each.
   Both parties please
Proceed without a breach
Of the peace to your lawful residence.'
Count what it costs these days to go to court,
And how little the families driven to that resort
Have left after expenses. It's the Law,
It's Perrin Dandin who eats up the rest –
   Who takes the wing and breast
And leaves the litigants the beak and claw.

9 Also *barrachois, barrisois.* Local pronunciation may be "barasway"
   (see chapter 4, note 2). A shallow estuary, lagoon, or harbour, of
   fresh or salt water, sheltered from the sea by a sandbar or low strip
   of land. *Dictionary of Newfoundland English (DNE),* 23.
10 Place Vendôme, Paris. These are allusions to people and objects
   that achieved prominence and notoriety during the events of
   1870–71: the Franco-Prussian War, the Siege of Paris, and the Commune. When the war began to go badly for France, Napoleon III
   abdicated and a new republican government was declared, which
   attempted to continue the war. By then, Paris was besieged by the
   Prussians. In October 1870, the Minister of the Interior in the new
   republican government, Léon Gambetta, made a celebrated escape
   from the city in a balloon, and after Paris surrendered, became the
   dominant force in the resistance to the Prussian invaders, which

was based in the city of Tours. Gambetta's efforts were in vain, and in March 1871 the republican leadership – by now deferring to Louis Adolphe Thiers – accepted defeat and agreed to an armistice. Thiers submitted to the demands and directions of Bismarck and the Prussians, even while trying to lead France into recovery. In Paris, however, the humiliation of the Prussian victory and the policies of the Thiers government led to the Paris Commune, which lasted from March until May. This was savagely crushed by the Thiers government; Paris suffered far more casualties and damage at the hands of the French army than it ever did from the Prussians. One symbolic casualty was the 155-foot Vendôme column, a bronze monument cast in 1810 from 1,200 captured Austrian and Russian cannon and topped by a statue of Napoleon Bonaparte in imperial regalia. Its destruction was presumably an act of vandalism against the discredited Second Empire and its legacy, the Thiers government. Thoulet's comment, that the Juillet column was "more fashionable these days," is perhaps an indication that at the time he was writing, republicanism was more politically correct than either monarchism or imperialism ("Juillet" presumably commemorates the storming of the Bastille on 14 July 1789). Certainly, in 1877, when Thiers died, his funeral was an excuse for massive displays of republicanism. By the mid-1880s, the stability of the Third Republic was practically assured. Thus, all of Thoulet's references here – to the Vendôme monument, the Juillet monument, Thiers, and Gambetta – are subtle yet significant observations on the politics of the day and on the displays of loyalty to republicanism in St-Pierre. His readers in France would thus have appreciated that as remote as St-Pierre might be from the French homeland, it was clearly and visibly as French as any community in France. I thank Olaf Janzen for clarifying this point and a number of others, especially those dealing with Newfoundland history and the French Shore.

11 Place de la Bastille, Paris.

12 Louis-Adolphe Thiers (1797–1877).

13 Léon Michel Gambetta (1838–1883).

14 Henriette Rosine Bernard (1844–1923), whose stage name was Sarah Bernhardt. "The golden voice," regarded as the greatest actress of her day. Although she toured the United States in 1881 (see Marie Colombier, *Le Voyage de Sarah Bernhardt en Amérique*), and England and South America in 1886, to my knowledge, she never visited St-Pierre and Miquelon.

15 Now known as Langlade.

16 In English in the original text.

17 Heroines in Sir Walter Scott's *The Pirate* (1821), set in the Orkney and Shetland Islands.

18 The insectivorous "pitcher plant," *Sarracenia purpurea*, official emblem of Newfoundland since 1954.

19 Jean-Jacques Rousseau (Geneva 1712–Ermenonville 1778). Writer

and philosopher. The Parc d'Ermenonville, which was inspired by Rousseau's vision of nature, includes a part known as *Le Désert,* and at one time included a number of *fabriques* or factory workshops. For details, visit membres.lycos.fr/parcsafabriques/erm/dErm1.htm/.

20 What Thoulet is describing here is, of course, that most distinctive of Newfoundland coastal shrubs, "tuckamore," which the DNE (586) defines as a "small stunted evergreen tree with gnarles spreading roots, forming closely matted ground cover on the barrens; (b) collectively, low stunted vegetation; scrub."

CHAPTER FOUR

1 Note by Thoulet: "The text that follows was written in 1886 and I do not wish to alter a single word."

2 The historical pronunciation of -ois was very close to *oé* or "way." Thus, the îles Groix became known as the Grey Islands, and *barachois* was often (and still is) pronounced "barrisway," in the same way that François, on Newfoundland's South Coast, was (and still is) pronounced locally as 'fran-sway.' The same vowel sound survives in certain francophone regions in Canada and elsewhere (southwestern Louisiana), where *moi* and *toi* are still pronounced *moé* and *toé.* See also DNE, 23.

3 Thoulet's spelling.

4 The River of Beans.

5 Cat Arm, Hare Bay, and Rabbit Bay.

6 Some early French maps show this as *Baie des désespoirs.* See note 2 regarding the local (and historically reasonably accurate) pronunciation of barachois as "*barasway.*" Perhaps the modern name Bay d'Espoir, with its local (and seemingly contradictory) pronunciation – Bay "Despair" – was actually not far from a reasonable approximation of the sixteenth-century French pronunciation of "désespoir,"' which would have sounded something like "dayzespware." This is somewhat convoluted, but may resemble how the two final syllables of the name would likely have been pronounced by the early French visitors to Newfoundland, besides being an accurate translation of the early French name (see Pope, *From Latin to Modern French,* 285). The likelier explanation connects the name shown on early maps, *Baie de désespoir,* with its literally translated meaning in the modern oral version, Bay "despair."

7 Deadmen's Island; Baie des trépassés – bay of the deceased.

8 Of England.

9 Thoulet makes no mention of spruce or pine, both of which were common on the west coast of Newfoundland.

10 It is worth noting that the moose, which is so closely identified with Newfoundland today, is not indigenous to Newfoundland and had

not yet been introduced to the island at the time Thoulet visited (that happened in 1904); hence his failure to mention it in his discussion.

11 In fact, mink, lynx, and squirrel are fairly common in various parts, especially of western Newfoundland, although Thoulet's brief stay obviously did not enable him to verify as much.

12 The very limited range of wildlife on Newfoundland perhaps contributed to the pattern of human extinctions there, regarding not only the Beothuk of historic times but earlier extinctions, which perhaps included those of the Maritime Archaic Indians and the Paleo-Eskimo. This theory, which is based on a biological analysis of the effects of limited flora and fauna, was developed by James Tuck and the late Ralph Pastore in "A Nice Place to Visit, but ...," which was itself inspired by an article by Bergerund, "Prey Switching in a Simple Ecosystem" (Tuck and Pastore, 69). The argument made by Tuck and Pastore has been challenged: by Frederick A. Schwartz in "Paleo-Eskimo and Recent Indian Subsistence and Settlement Patterns on the Island of Newfoundland," and by Priscilla Renouf in "Prehistory of Newfoundland Hunter-Gatherers: Extinctions or Adaptations?"

13 "It cannot continue much longer," and of course it did not. The French Shore question would be settled in 1904. See Hiller, "The Newfoundland Fisheries Issue," mentioned earlier, and "The 1904 Anglo-French Newfoundland Fisheries Convention," 82–98.

14 Thoulet is clearly no historian. He is incorrect in his statement that Gilbert brought 250 colonists to Newfoundland in 1583; Gilbert's personnel were mostly adventurers, and visited St John's in 1583 while returning to England, having failed to reach their intended destination – namely, continental North America. The stopover in Newfoundland was very much an opportunistic affair; see David Quinn, *Sir Humphrey Gilbert and Newfoundland* (reprinted in a collection of Quinn's writings, *Explorers and Colonies: America, 1500–1625,* 207–24). As well, Quinn wrote the essay on Gilbert that appears in the *Dictionary of Canadian Biography*, vol. 1.

15 Fishermen may well have been wintering over at Plaisance [Placentia] earlier in the 1600s. That said, the first attempt to establish a permanent outpost at Plaisance was not made until 1655, and it was not until 1662 that a governor arrived with thirty soldiers, eighteen cannon for the defences, and settlers.

16 Jean-Baptiste Marquis de Seigneley (1619–83). French statesman and financier. A good source on Colbert and French efforts to colonize Newfoundland is Laurier Turgeon, "Colbert et la pêche française à Terre-Neuve."

17 Thoulet gets the date wrong here. The Treaty of Aix-la-Chapelle, signed in 1748, ended the War of the Austrian Succession. France and England did not formally become belligerents in that war until 1744, well after the date Thoulet gives for the treaty.

18 For an analysis of the complex process by which the eastern limit of the French Shore was shifted from Cape Bonavista to Cape St John, and the western limit was shifted from Point Riche to Cape Ray, see Jean-François Brière, "Pêche et politique." The French navy's successful recovery of its strength and prestige between 1763 and 1783 is the focus of Jonathan Dull, *The French Navy and American Independence.* The diplomacy through which French rights in Newfoundland were reasserted and defined in 1783 is paid careful attention by Orville T. Murphy in "The Comte de Vergennes." Murphy eventually wrote the definitive biography of Vergennes: *Charles Gravier, Comte de Vergennes.*

19 Étienne François, duke of Choiseul (Nancy 1719–Paris 1785). French statesman, minister, and diplomat, and protégé of Madame de Pompadour. Signed the Treaty of Paris in 1763, which ended the Seven Years' War. *Petit Larousse Illustré*, 1999.

20 Louis Gilbert Guillouet d'Orvilliers (1757–86). Garde de la marine (1771); naval lieutenant (1782); captain of *Le Malin* when she went down in November 1786. *Nouveau Larousse Illustré.*

21 Charles Henri, Count d'Estaing (Ravel, Puy-de-Dôme, 1729–Paris 1794). French admiral, decorated in the War of American Independence. Commanded the Garde Nationale at Versailles in 1789. Died by guillotine. *Petit Larousse Illustré*, 1999.

22 François Joseph Paul, Count de Grasse (Le Bar, Provence, 1722–Paris 1788). French naval officer during the American War of Independence. *Petit Larousse Illustré*, 1999.

23 French naval admiral, Bailli de Suffren. The most recent biography of this famous and highly successful French naval commander is Cavaliero, *Admiral Satan.* www.geneastar.org/fr/bio/.

24 Charles Gravier, Count of Vergennes (Dijon 1719–Versailles 1787). French statesman and diplomat. Reaffirmed France's prestige after the Seven Years' War. Contributed to the independence of the United States in 1783. Signed a treaty enabling commerce with England in 1786. *Petit Larousse Illustré*, 1999.

25 St-Pierre and Miquelon were captured by British forces in 1793, not 1798, as Thoulet states. See Hitsman, "The Capture of St. Pierre-et-Miquelon, 1793," 77–81.

26 Thoulet is, of course, oversimplifying and to an extent distorting the true picture, although to his credit, he is not far wrong. Newfoundland did not declare its independence in 1855; it was granted "responsible government," which – at the risk of oversimplifying the meaning of the term – conceded responsibility for local self-government to Newfoundland (this, a few years after the same privilege was extended to other British American colonies). This hardly constituted "independence," since England retained key powers over diplomacy, war, and so on, and could still intervene and disallow measures passed by the Newfoundland government that might conflict with the interests of the mother country or of the empire.

So it was entirely appropriate for France to ignore the Newfoundland Assembly and deal directly with the British government in London; indeed, it is highly unlikely that had France attempted to deal directly with Newfoundland, England would have tolerated so blatant a violation of diplomatic protocol. It is likely that the Newfoundland colonial government would have been highly embarrassed as well. The abortive attempt to resolve the French Shore question in 1857, which stirred up such emotional heat in Newfoundland, is discussed in Neary, "The French and American Shore Questions," 95–122. Se also R.A. MacKay, "Responsible Government and External Affairs," 265–74.

27 See previous note.

28 Joseph Arthur de Gobineau. See Wilkshire, *A Gentleman in the Outports.*

29 Newfoundland joined Confederation in April 1949.

30 Many readers will of course know that most Newfoundlanders are descended from either Irish or West Country English settlers. That said, a small number of Scottish Highlanders did settle on Newfoundland's West Coast in the nineteenth century, around Jeffreys and St David's; see Rosemary Ommer, "Highland Scots Migration to Southwestern Newfoundland: A Study of Kinship," 212–33; see also Ommer's MA thesis, from which that article was derived.

CHAPTER FIVE

1 Woody Point.

2 Thoulet is describing the ultramafic mantle rocks, primarily peridotites, found in the region now known as the Tablelands in Gros Morne National Park, designated by UNESCO as a World Heritage Site.

3 Likely Lieutenant Louis Koenig. See Introduction, page xxiv.

4 Port Saunders.

5 A prize named after the Baron of Monthyon, who in 1819 donated 12,000 francs to the French Academy in order to revive the "good deeds of the people" and to inspire new ones. During the Monarchy of July, 351 people (of which 216 were women) were judged worthy of the prize. See *Revue d'histoire du XIX siècle,* no. 10, 1994.

6 *Qui veut la fin veut les moyens.* The end justifies the means.

CHAPTER SIX

1 "A late nineteenth century, post-impressionist school of painting that emphasized light and the contrast between shadow and light." *Trésor de la langue française.*

2 The Tearful One.

3 Dutch/French academic painter (1795–1858). His work can be

found in the following museums and art galleries: National Gallery, London; Hermitage, Saint Petersburg; Wallace Collection, London; Van Gogh Museum, Amsterdam; Cleveland Museum of Art, Ohio; Joconde Database of French Museum Collections (in French); National Gallery of Victoria, Australia; National Portrait Gallery, London.

4 Jean-Jacques Rousseau, *Confessions,* Book VII (1743–44), part 2. The reference is to the Venetian courtesan Zulietta, who in a moment of anger at Rousseau pushes him away, declaring, "Zanetto, lascia le donne, e studia la matimatica" (Jean, leave the women, study mathematics instead). Thoulet seems to have gotten her "advice" wrong.

5 Green, Woody, Brush, Flat and Bird Islands.

CHAPTER SEVEN

1 Large, mythical sea monster said to appear off the coast of Norway.

2 Thoulet indicated earlier that he was aware that the Norse perhaps settled briefly in Newfoundland (see his remarks on page 44). It is interesting that had he turned to his right while rounding Cape Onion, instead of left, he would have seen L'Anse aux Meadows, the site (unbeknownst to him) of a short-lived Norse settlement,.

3 Brittany, France.

4 Another description of tuckamore.

5 A small, rounded hillock. *Oxford Dictionary & Thesaurus.*

6 Tityrus: "Giant. Son of Gaia (Earth) who was killed for assaulting the goddess Leto, by Zeus or by Apollo. Odysseus (Od. Book I) saw him lying in the underworld, covering nine plethra ('acres') of ground, while two vultures tore at his liver." M.C. Howaltson, *Oxford Companion to Classical Literature,* 657.

7 G. Chaucer, *The Tale of Melibee and Dame Prudence.*

8 Genille Point (Genile Point, Point Genille). See Fishermans Cove (Brig Bay, Anse des Pêcheurs). Seary, *Place Names of the Northern Peninsula,* 97.

9 Probably "Patrick Kearney," *Journal of Legislative Council of the Island of Newfoundland. Appendix – List of Inhabitants on the French Shore, Newfoundland,* St John's, 1873, 486.

10 The pitcher plant. See 173n18 above.

11 Capitaine Félix Auguste Le Clerc. See 167n45, note 45 above.

12 Perhaps to calculate the exact time.

13 Thoulet, "Sept mois chez les Chippeways," 601, and "Le Territoire du Montana et le Parc national des Etats-Unis," 721.

14 Thoulet, a European, refers to them as elk. We would, of course, call them moose.

15 "When one is torn away, another succeeds." Virgil.

CHAPTER EIGHT

1   Or Noddy Bay. Seary, *Place Names of the Northern Peninsula,* 123.
2   Koenig (394): "The name [shoe cutter] is a good indication of
    what the ground is like. In fact, it is made up entirely of schist, with
    a few cubes of iron pyrite scattered here and there."
3   Wood Cove.
4   Cod fish split, salted but not dried. *DNE,* 224.
5   The cod trap was developed in the 1860s. See
    www.stemnet.nf.ca/cod/history7.htm/.
6   Baie aux mauves. Mauve Bay, See Noddy Bay, Noddy Harbour.
    Seary, *Place Names of the Northern Peninsula,* 127.
7   Anonymous, "The Children in the Wood," in Arthur Quiller-Couch,
    Arthur, ed. (1863–1944). *Oxford Book of Ballads,* 1910.
8   Bad Lad.
9   Vice-Admiral Georges-Charles Cloué (1817–1889). Known for his
    hydrographic work on Newfoundland's coasts. See his *Pilote de Terre-
    Neuve.* See Tréfeu, *Nos marins,* 102–8. See Fonds Paul-Émile Miot
    for a photograph of Capitaine Georges-Charles Cloué à bord du
    navire l'ARDENT 1857 Album. Colour transparency available/Trans-
    parent couleur disponible Numbers: Accession: 1995-084 Repro-
    duction: PA-194627.
    www.archives.ca/05/050402/05040203_e.html/.
10 See Fonds Paul-Émile Miot (Lieutenant aboard the *Ardent*). Rocher
    peint par les marins français 1857–1859/Baie du Sacre, Terre-
    Neuve –1 item–21.5 x 28.4 cm Reference numbers: Accession
    Reproduction: -PA-188210 (copy negative number) Paul-Émile
    Miot/Archives nationales du Canada/PA-188210.
    www.archives.ca/05/050402/05040203_e.html
11 Warbler's or sparrow's.
12 Probably Koenig. See page xxiv.

CHAPTER NINE

1   The Norse did not venture into the North Atlantic in their *drakkars*
    or "dragon-boats," as Thoulet asserts. Although some recent recon-
    structions of Norse longships (to use the more correct term) have
    been sailed across the Atlantic, in fact the Norse did not use long-
    ships for oceanic travel; they were more suited to the rivers and
    coastal waters of Europe. The vessel of choice for the open sea was
    the *knarr,* a vessel that was beamier, deeper, and more weatherly.
    See, for instance, Unger, "The Archaeology of Boats: Ships of the
    Vikings," 20–7; Roald Morcken, "Longships, Knarrs and Cogs,"
    391–400; Christenson, "Viking Age Ships and Shipbuilding,"
    19–28; and Vinner, "Unnasigling – the seaworthiness of the mer-
    chant vessel," 95–108.

2 Thoulet places "Vinland" in the United States. Although it is possible to make a case for a location south of Nova Scotia, the conventional view today is that Vinland was somewhere in the Gulf of St Lawrence. This would be consistent with the sailing times as well as the flora and fauna described in the sagas, supported by archaeological findings at L'Anse aux Meadows.

3 There is today no certainty about the factors leading to the Norse expansion. It would probably be unwise to assume a single explanation, such as famine. A reliable source is Sawyer, "Scandinavia in the Viking Age," 27–30. See also his earlier work, *Kings and Vikings; Scandinavia and Europe* AD *700–1100* (London: Methuen, 1982).

4 Thoulet's remarks, including "without codfish, Newfoundland would have remained uninhabited for a long time," are of course Eurocentric, and ignore the fact that Newfoundland was very much inhabited by indigenous people long before Europeans arrived to fish for cod. It also ignores the fact that those same indigenous people were able to subsist quite successfully without cod, by harvesting a broad range of marine resources, including seals, salmon, and capelin. They did not rely on cod because they did not have the means to fish offshore for groundfish.

5 Thoulet's choice of language here in describing how and where France fished in Newfoundland waters "since the Treaty of Utrecht" obscures the fact that the limits of the treaty shore as defined by the Treaty of Utrecht were modified in 1783. After that year the treaty shore limits were Cape St John in the east to Cape Ray in the west. Thoulet's statement here makes reference only to the revised post-1783 limits, although his chronological "start point" is 1713, when the Treaty of Utrecht was signed.

6 A vessel engaged in the cod fishing on the Newfoundland offshore grounds, especially the Grand Banks. DNE, 21.

7 In the cod fishery, a long, buoyed fishing line with closely placed and baited hooks attached at intervals; set-line; trawl. DNE, 73.

8 "Member of a fishing crew who cuts around the backbone of codfish brought ashore to be dressed, opening the fish for salting and drying." DNE, 510.

9 "Part of a fish near the head, synonym : ears or collar". Cf. *chignon, chignon du cou.* Brasseur et Chauveau, *Dictionnaire des régionalismes de Saint-Pierre et Miquelon,* 191–2.

10 "Member of a fishing crew who applies salt in the process of processing dried cod." DNE, 432.

11 DNE, 375.

12 Local spelling. Sometimes written *houari.* Borrowed from the English "wherry."

13 Here Thoulet states: "To write a history of the fishery, going back if possible to the first ships that were sent out to fish for cod, especially under Louis XIV and Louis XV." To the unwary reader, this might suggest that the first French fishing ships to sail regularly to

Newfoundland did so during the reigns of Louis XIV and XV – that is, in the late seventeenth and early eighteenth centuries. In fact, the first French fishermen appeared off Newfoundland in 1504 (as Thoulet himself had already pointed out on page 44), and by the mid-seventeenth century, French fishing vessels outnumbered English ones by roughly two to one; see Laurier Turgeon, "Colbert et la pêche française à Terre-Neuve," 255. Though some argue that the high point of the French fishery appears to have been reached between 1678 and 1688, which would make Thoulet correct in his statement, there is also evidence that the French fishery had already gone into a decline by then (there was a 20 per cent drop in the size of the French fishing fleet between 1664 and 1686); see Laurier Turgeon, "Le temps des pêches lointaines. Permanences et transformations (ca 1500 – ca 1850)," 133–81.

14 Samuel de Champlain (Brouage 1570–Québec 1635), French explorer, cartographer, Governor of New France; Louis de Buade, Count de Frontenac (St-Germain, 1622–Québec, 1698), Governor General of New France; Cavalier de la Salle (Rouen 1643–Texas 1687), explorer of Louisiana and the Mississippi; Louis Jolliet ou Joliet (Baptized at Québec 1645–Canada 1700), explorer, cartographer, fur trader, and the King's Hydrographer, discoverer of the Mississippi River; Pierre Lemoyne d'Iberville et d'Ardillières Céleron (Baptized at Montreal 1661–Havana[?] 1706); Louis Joseph Montcalm, Marquis de St-Véran (Candiac 1712–Québec 1759), general, killed in the Battle of the Plains of Abraham. All of these individuals have been provided with excellent biographical essays in the *Dictionary of Canadian Biography*.

15 Molière, *Le dépit amoureux* V iv. The conventional, stereotype stock character Gros-René also appears in *Le Médecin volant* and S*ganarelle*.

16 See Magord, *Une minorité francophone hors Québec : les Franco-Terreneuviens*, for a recent, in-depth study of "this micro-society that is showing all the undeniable signs of a renaissance" – J.M.Lacroix, *Préface*.

17 Ange Duquesne de Menneville (Toulon 1700–Antony 1778), naval officer and Governor General of New France from 1752 to 1755; Jean Bart (1650-1702), privateer from Dunkerque in the service of Louis XIV of France; Anne-Hilarion de Cotentin de Tourville, comte de Tourville (Paris 1642–Paris 1701), Knight of the Order of Malta, Maréchal de France, officer of the *Marine Royale*; René Duguay-Trouin (St-Malo 1673–Paris 1736), real name René Trouin, sieur du Gué, also known as "du Guay-Trouin," privateer under Louis XIV.

18 The etymology of bait according to the OED (1, 890 c.) is: "ON beit, pasture. beita (fem.) food to entice a prey. Cogn. In OE bát f. food. MHG beiz n. beiz f. hunting."

19 Vietnam.

CHAPTER TEN

1 Gustave Doré (1832–1883). Artist, engraver, and illustrator of the Bible, Rabelais, La Fontaine, Dante, Cervantès, and seventeenth-century fairy tales. lescontesdefees.free.fr/images/galerie_des_gravures_de_gustave.htm/.
2 Or Canada Harbour.
3 Or Wild Cove.
4 "The line in the bottom of a valley in which the slopes meet and which forms a natural water course; also, the line which follows the deepest part of a bed or channel of a river or lake." (germ.: bottom of a valley) *OED*.
5 "I know not what trifles it is thinking about."
6 "Nature does not proceed by leaps." – Linnaeus, Carl von Linné (1707–78), *Philosophia Botanica*, 1750. In J. Bartlett, *Familiar Quotations. A Collection of passages, phrases and proverbs traced to their sources in ancient and modern literature.* An adage often used by Darwin in *The Origin of Species* (1859).

CHAPTER ELEVEN

1 Tête-de-Mort.
2 Seary, *Place Names of the Northern Peninsula,* 43.
3 *Outarde* or *bernache du Canada,* 'Canada goose' in English, not to be confused with Goose Bay, Labrador. Also Anse aux outardes, or Bustard Cove. Seary, *Place Names of the Northern Peninsula,* 59, 134.
4 The origins of the name "Labrador" are in fact Portuguese, not Spanish; see entries for "Labrador" in the *Encyclopedia of Newfoundland and Labrador* and in the *DNE*.
5 Honoré (de) Balzac (1799–1850). French novelist, author of *La Comédie Humaine.*
6 "New land," an island in the Russian Arctic.
7 "Ingonish : name used by Champlain as 'Niganis' and by Nicholas Denys as 'Niganiche' may be of Portuguese, but is more probably of Micmac origin." *Place Names and Places of Nova Scotia.*
8 "Of colours, it is good to hold green, red, yellow and white, and by all means to have light enough with windows in the day, wax candles in the night." R. Burton, *The Anatomy of Melancholy,* part 2, section 2, 66.
9 Italian soprano (1843–1919) who toured Europe and the United States: "superbly controlled voice of wide range, perfect evenness and extraordinary flexibility." H. Weinstock, co-auth. *Men of Music,* quoted in *Encyclopaedia Americana,* 540.
10 Swedish soprano (1843–1921) who toured Europe and the United States: "noted for the lyrical sweetness and brilliance of her voice." *Encyclopaedia Americana,* 355.

11 Molière, *L'amour médecin* (1665) II, vii.

12 Thoulet, "Sept mois chez les Chippeways," 601.

13 A liniment used for sprains or bruises.

14 *Extrait de saturne.* Acétate basique de plomb liquide. Goulard's extract, once used as a lotion. The chemical name in English is "a solution of basic lead (II) acetate." Louis de Vries, *French–English Science Dictionary.*

15 "Let them be as they are or let them not be."

16 Cantal. A department in south-central France (capital city, Aurillac). It includes a portion of the Massif Central, a volcanic, mountainous region. Its economy is largely agricultural (dairy cows and cheese production).

17 Baron Georges Cuvier (Montbéliard 1769–Paris 1832). Zoologist and palaeontologist who founded comparative anatomy and vertebrate palaeontology.

18 Della Robbia. Sculptors and ceramists. "A Florentine family of sculptors and ceramists active from the 15th to the 16th century. The most famous is Luca (Florence, 13??–1482). He studied the work of Ghiberti and Nanni di Banco and also learned a great deal from Donatello. Apart from the excellent quality of his sculpture, he was also famous for his invention of glazed terracotta, whose secret was passed down to his nephew Andrea (Florence 1435-1525) and to Andrea's son, Giovanni (Florence 1469–1529)." www.mega.it/eng/egui/hogui.htm/.

19 In English in the original text.

CHAPTER TWELVE

1 The place that originally inspired the name Twillingate in Notre Dame Bay, Newfoundland, as many readers will recognize.

Bibliography

MANUSCRIPT SOURCES

Archives de l'Académie des sciences, Paris
Archives, Collège de France, Paris
Archives du Musée océanographique de Monaco
Fonds des colonies, Archives nationales de France, Paris
Bibliothèque du Muséum d'histoire naturelle, Paris
Service historique de la Marine, Vincennes
Musée historique de la marine, Paris
Musée Archives L'Arche, Saint-Pierre et Miquelon
Provincial Archives of Newfoundland and Labrador
Centre for Newfoundland Studies, QEII Library, Memorial University
    of Newfoundland.

PRINTED SOURCES

Andrieux, Jean-Pierre. *St. Pierre and Miquelon: A Fragment of France in
    North America.* Lincoln, Ont.: W.F. Rannie 1983
Annales de physique et chimie. 1880. vol. 20.
Annuaire de la Marine et des Colonies. 1986. Paris: Berger-Levrault 1887
Auerbach, Bertrand. *Bulletin trimestriel de la Société de géographie de l'est.*
    4e trimestre, 1913. 316

Barnhart, C., and W. Halsay, eds. *The New Century Cyclopedia of Names.*
    3 vols. New York: Appleton Century Crofts 1954
Bartlett, J. *Familiar Quotations. A Collection of passages, phrases and proverbs
    traced to their sources in ancient and modern literature.* 16th ed. Boston,
    Toronto, and London: Little, Brown 1992
Bernet, Étienne. *Bibliographie francophone de la grande pêche. Terre-
    Neuve–Islande–Groenland.* Fécamp: Musée des terre-neuvas et de la
    pêche 1998

Birnhaum, Pierre. *La France imaginée.* Paris: Gallimard Folio/Histoire 2003

Blais, Suzelle. *Apport de la toponymie ancienne aux études sur le français québécois et nord-américain, Documents cartographiques du régime français.* Études et recherches toponymiques 6. Québec: Commission de toponymie, Gouvernement du Québec 1983

Bourgin, Georges. *Inventaire des archives de la marine, service hydrographique, sous-série JJ (Journaux de bord) déposée aus Archives nationales. Revu complété par* Etienne Taillemite. Paris: Imprimerie nationale 1963

Brasseur, P., and J.-P. Chauveau. *Dictionnaire des régionalismes de Saint-Pierre et Miquelon.* Tübingen: Max Niemeyer Verlag 1990

Brière, Jean-François. "Pêche et politique – Terre-Neuve au XVIIIe siècle: la France véritable gagnante du traité d'Utrecht?" *Canadian Historical Review* 64, no. 2 (June 1983): 168–87

Brown, George W., ed. *Dictionary of Canadian Biography.* 24 vols. Toronto: University of Toronto Press and Les Presses de l'université Laval 1979

Bulletin de la Société de Géographie, vol. 13 (1867): 610–11

Bulletin trimestrial de la société de géographie de l'Est, 4e trimestre (1913): 316

Burton, R. *The Anatomy of Melancholy.* New York: Vintage 1977

Carpine, Christian. *La pratique de l'océanographie au temps du Prince Albert Ier.* Monaco: Musée Océanographique 2002

Carpine-Lancre, Jacqueline. "Chronology of the Main Events Related to the Origins, and the First and Second Editions of 'La Carte bathymétrique des oceans.'" *Charting the Secret World of the Ocean Floor. The GEBCO Project, 1903–2003.* Collected presentations at the GEBCO Centenary Conference, Monaco, 14–16 April. International Hydrographic Bureau, Monaco. 2003a (CD-ROM). www.ngdc.noaa.gov/mgg/gebco/gebco.html

– "Les croisières océanographiques françaises antérieures à 1914." XIIe Congrès international d'histoire des sciences Paris. 25–31 août 1968. Monaco: Musée océanographique 1968

– "Les expéditions océanographiques françaises du XIXe siècle." Actes du *12e Congrès international d'histoire des sciences.* Paris. vol. 7 (1971): 61–5

– "Julien Thoulet (1843–1936) Pionnier de l'Océanographie française." *Abstracts 18th International Congress of the History of Science, 1–9 August 1989.* eds. Fritz Kraft and Christophe J. Scriba. Hambourg-Munich, R8/3 (1989) 1

– Unpublished letter to the translator. 9 April 2000

– "Origins of a Lasting Bathymetric Endeavour." *International Hydrographic Review* 4, no. 2. 2003b (August): 6–16

– "Corrigenda." *International Hydrographic Review* 4, no. 3 (New Series). 2003c (December): 96

– Unpublished letter to the translator. 24 March 2004.

– "The Origin and Early History of 'La Carte générale bathymétrique des oceans.'" *History of GEBCO 1903–2003.* 15-51. Lemmer: Geomatics and Information Trading Centre

Catalogue des appareils d'océanographie en collection au Musée océanographique de Monaco (Monaco) 73, no. 1437: 50–3

Cavaliero, Roderick. *Admiral Satan: The Life and Campaigns of Suffren.* London: I.B. Tauris 1994

Cazeils, Nelson. *Cinq siècles de pêche à la morue. Terre-Neuvas & Islandais.* Rennes: Éditions Ouest-France 1997

Christenson, Arne Emil. "Viking Age Ships and Shipbuilding." *Norwegian Archaeological Review* 15, nos. 1–2 (1982): 19–28

Cloué, Georges-Charles (Vice–Admiral). *Pilote de Terre-Neuve.* Paris: Lainé 1869 and Paris: Imprimerie nationale 1882

Deacon, Margaret. *Scientists and the Sea 1650–1900: A Study of Marine Science.* London: Ashgate 1997

Dérible, Marc. *Mos et expressions de Saint-Pierre-et-Miquelon.* Saint-Pierre: Imprimerie Administrative 1986

De Vries, Louis. *French-English Science Dictionary for Students in Agricultural, Biological and Physical Sciences.* New York and London: McGraw-Hill 1940

Dull, Jonathan. *The French Navy and American Independence – A Study of Arms and Diplomacy 1774–1787.* Princeton, N.J.: Princeton University Press 1975

Encyclopaedia Americana. vol. 21. Danbury, Ct.: Grolier 2000.

Encyclopedia of Newfoundland and Labrador. 5 vols. St John's: Newfoundland Book Publishers 1981–94

Encyclopédie du Canada. Montreal: Stanké 2000

Fiero, Alfred. *Inventaire des photographies sur papier de la Société de Géographie.* Paris: Bibliothèque nationale, Département des cartes et plans 1986

Filhol, Henri. *La vie au fond des mers. Les explorations sous-marines et les voyages du Travailleur et du Talisman.* Paris: G. Masson 1885

Folin, Léopold de. *Sous les mers. Campagnes d'exploration du "Travaiailleur" et du "Talisman."* Paris: J.-B. Baillère et fils 1887

France, Peter. *The New Oxford Companion to Literature in French.* Oxford: Clarendon Press 1995

Gonin, J.C. *Dictionnaire anglais–français des termes nautiques English–French Dictionary of Nautical Terms.* St-Malo: Editions A Compte d'auteur 2000

Grossetête, Abbé J.-M. *La grande pêche de Terre-Neuve et d'Islande.* St-Malo: Éditions L'Ancre de marine 1988. First edition 1921

Guide des termes de marine, Petit dictionnaire thématique de marine. Douarnanez: Le chassé-Marée Aren 1997

Guilbert, Louis, et al. *Grand Larousse de la langue française en sept volumes.* Paris: Larousse 1986

Guyotjeannin, Olivier. *Saint-Pierre et Miquelon*. Paris: Editions l'Harmattan 1986

Hiller, James. "Utrecht Revisited: The Origins of French Fishing Rights in Newfoundland Waters." *Newfoundland Studies* 7, no. 1 (Spring 1991): 23–39.
–"The 1904 Anglo-French Newfoundland Fisheries Convention: Another Look." *Acadiensis* 25, no. 1 (Autumn 1995): 82–98
–"The Newfoundland Fisheries Issue in Anglo-French Treaties, 1713–1904." *Journal of Imperial and Commonwealth History* 24, no. 1 (January 1996): 1–23
Hitsman, J. Mackay. "The Capture of St. Pierre-et-Miquelon, 1793." *Canadian Army Journal* 13, no. 3 (July 1959): 77–81
Howaltson, M.C., and Ian Chilvers, eds. *The Concise Oxford Companion to Classical Literature*. Oxford and New York: Oxford University Press 1993

Imbs, Paul, dir. *Trésor de la langue française: dictionnaire de la langue du XIXe et du XXe siècle (1789–1960)*. 16 vols. Paris: Editions du Centre national de la recherche scientifiques et Gallimard 1971–

Journal of Legislative Council of the Island of Newfoundland. St John's: J.C. Withers 1873
Journal of the House of Assembly 1872. Appendix Fisheries. Tabular Statement of Statistics on the Newfoundland Coast. 633–71, French Shore

Kirwan, William. "Selected French and English Fisheries Synonyms in Newfoundland." *Regional Language Studies in Newfoundland* 9 (1980): 10–21
Knowles, Elizabeth, editor. *Oxford Dictionary of Quotations*. 5th ed. Oxford: Oxford University Press 1999
Koenig, Louis. "Le 'French Shore' (souvenirs de campagne à Terre-Neuve)." *Tour du monde* 60 (1890): 369–400

Lebailly, Andrée. *Saint-Pierre et Miquelon Histoire de l'archipel et de sa population*. St-Pierre-et-Miquelon: Editions Jean-Jacques Oliviero 1988
Lefort, A., and L. Lemesle. *La Royale et les Terre-Neuvas*. St-Malo: L'Ancre de Marine 1994
"Le Gulf-Stream par le lieutenant J.E. Pillsbury U.S.N." (traduit des Pilot-Charts, aout 1894). *Bulletin de la Société de géographie de Marseille* 19, no. 3 (1895): 269
" Lettre du 28 octobre 1886 au Vice-amiral en chef, Préfet maritime à Brest par le Directeur chargé de la construction navale. " *Nouvelles de la flotte. Correspondance générale. Lettres envoyées. Cherbourg. Brest.* Vincennes : Service historique de la marine BB/2/640
MacKay, R.A. "Responsible Government and External Affairs." In R.A.

MacKay, ed., *Newfoundland: Economic, Diplomatic and Strategic Studies.* Toronto: Oxford University Press 1946. 265–74

Magord, André. *Une minorité francophone hors Québec: les Franco-Terreneuviens.* Tübingen: Max Niemeyer Verlag 1995

Martin, Jean-Pierre. *Rue des Terre Neuvas. Normands & Bretons à Terre-Neuve au XIXème siècle.* Rouen: Les éditions du Veilleur de Proue 2001

Maury, M.F. *Steam-lanes across the Atlantic.* Washington: U.S. Government Printing Office 1873. First published 1855

Mollat, Michel, ed. *Histoire des pêches maritimes en France.* Toulouse: Privat 1987

Morandière, Charles de la. *Histoire de la pêche française dans l'Amérique septentrionale.* 3 vols. Paris: Maisonneuve et Larose 1962–66

Morcken, Roald. "Longships, Knarrs and Cogs." *Mariner's Mirror* 74, no. 4 (November 1988): 391–400

Murphy, Orville T. "The Comte de Vergennes, the Newfoundland Fisheries, and the Peace Negotiation of 1783: A Reconsideration." *Canadian Historical Review* 46, no. 1 (March 1965): 32–46

Neary, Peter. "The French and American Shore Questions as Factors in Newfoundland History." In James K. Hiller and Peter Neary, eds., *Newfoundland in the Nineteenth and Twentieth Centuries: Essays in Interpretation.* Toronto: University of Toronto Press 1980. 95–122

Ommer, Rosemary. "Highland Scots Migration to Southwestern Newfoundland: A Study of Kinship." In John Mannion, ed., *The Peopling of Newfoundland.* St John's: ISER 1978. 212–33
– *Scots Kinship, Migration and Early Settlement in Southwestern Newfoundland.* MA thesis, Memorial University of Newfoundland 1974

Peau, Etienne. "Monsieur le Professeur Julien Thoulet, Patriarche de l'Océanographie." *Bulletin de la Société des Amis de l'Institut Océanographique du Havre.* 18e année, 52, (1936): 7

Perret, Robert. *La géographie de Terre-Neuve.* Paris: Guilmoto 1913

Le Petit Larousse Illustré. Paris: Larousse 1999

Place Names and Places of Nova Scotia. Public Archives of Nova Scotia. Belleville, Ont.: Mika 1982

Pope, M.K. *From Latin to Modern French.* Manchester: Manchester University Press 1934

Portier, P. "Le professeur J. Thoulet. Discours prononcé le 4 janvier 1936." *Bulletin de la Société d'Océanographie de France (Anciennement du Golf de Gascogne)* 16e Année, 90 (15 Juillet 1936): 1553–4

Querré, Christian. *La grande aventure des Terre-Neuvas de la baie de Saint-Brieuc.* St-Brieue: Éditions du Dahin 1998

Quiller-Couch, Arthur, ed. *The Oxford Book of Ballads.* Oxford: Clarendon 1910

Quinn, David. *Sir Humphrey Gilbert and Newfoundland.* St John's: New-

foundland Historical Society 1983. Reprinted in David Quinn. *Explorers and Colonies: America, 1500–1625.* London and Ronceverte, WV: Hambledon Press 1990. 207–24
– "Sir Humphrey Gilbert." *Dictionary of Canadian Biography.* Québec et Toronto : Presses de l'Université Laval et University of Toronto Press. vol. 1: 331–6.

Renouf, Priscilla. "Prehistory of Newfoundland Hunter-Gatherers: Extinctions or Adaptations?" *World Archaeology* 30, no. 3 (1999): 403–20
Rompkey, Ronald. *Anthologie des voyagers français 1814–1914,* Rennes: Presses Universitaires de Rennes 2004
[Rouch, Jules]. "Le Prince Albert et le Professeur Thoulet." *Bulletin trimestriel des Amis du Musée océanographique de Monaco* 22 (1952): 1–5
[–] "Le Prince Albert et le Professeur Thoulet, note complémentaire." *Bulletin trimestriel des Amis du Musée océanographique de Monaco* 28 (1953): 10–11
Rousseau, Jean-Jacques. *Œuvres Complètes I. Les confessions.* 2e Partie, Livre 7. Paris: Gallimard bibliothèque de la Pléiade 1979: 322

Salkin-Laparra, Geneviève. *Marins et diplomats. Les attachés navals 1860–1914.* Vincennes: Service Historique de la Marine 1990
Sawyer, Peter H. "Scandinavia in the Viking Age." In William Fitzhugh and Elizabeth Ward, eds. *Vikings: The North Atlantic Saga.* Washington, D.C.: Smithsonian Institution Press 2000. 27–30
Schwartz, Frederick A. "Paleo-Eskimo and Recent Indian Subsistence and Settlement Patterns on the Island of Newfoundland." *Northeast Anthropology* 47 (Spring 1994): 55–70
Seary, E.R. *Place Names of the Northern Peninsula.* R. Hollett and William Kirwin, eds. St John's: ISER Books 2000
Secrétariat d'État (Secretary of State Canada). *Bateaux et Engins et pêche* (Boats and Fishing Gear). *Bulletin de terminologie* (Terminology Bulletin) 158. Ottawa: Ministre des Approvisionnements et Services (Supply and Services Canada) 1977
Secrétariat d'État (Secretary of State Canada). *Flore du Canada* (Canadian Flora). Bulletin de terminologie (Terminology Bulletin) 156. Ottawa: Ministre des Approvisionnements et Services (Supply and Services Canada) 1974
Simpson, J.A., and E.S.C. Weiner, eds. *Oxford English Dictionary.* 20 vols. Oxford: Clarendon Press 1989
Story, G.M., W.J. Kirwin, and J.D.A. Widdowson. *Dictionary of Newfoundland English.* Toronto: University of Toronto Press 1982

Thomson, C. Wyville, ed. *Report on the Scientific Results of H.M.S. Challenger during the Years 1873–76;* Zoology. Vol. I. London : HMSO 1880
Thoulet, Julien. "Sept mois chez les Chippeways." *Revue scientifique de la France et de l'étranger.* 2e série, 26 (1873): 601
– "Le Territoire du Montana et le Parc national des Etats-Unis." *Revue*

scientifique de la France et de l'Étranger. 2e série, 31 (1874): 721

– *Contributions à l'étude des propriétés physiques et chimiques des mineraus microscopiques.* Diss. U. of Paris 1880

– "Sur les spicules d'éponges silicieuses vivantes recueillies par le *Talis-man*." *Comptes Rendus de l'Académie des Sciences* 98 (1881): 1000–1. Note de J. Thoulet présentée par A. Milne-Edwards et *Bulletin de la Société de minéralogie de France* 7 (1881): 147

– "Expériences relatives à la vitesse des courants d'eau ou d'air suscepti-bles de maintenir des grains minéraux de volume et de densité déter-minés." *Comptes rendus de l'Académie des sciences* 107 (1884)

– "Attraction s'exerçant entre deux corps en dissolution et les corps solides immergés." (1ere partie: 1885a; 2e partie: 1885b). *Comptes rendus de l'Académie des sciences* 100

– "Sur le mode d'érosion des roches par l'action combinée de la mer et de la gelé." *Comptes rendus de l'Académie des Sciences* 103 (1886a): 1193

– "Sur le mode de formation des bancs de Terre-Neuve." *Comptes rendus de l'Académie des Sciences* 103 (1886b): 1042

– "Observations faites à Terre-Neuve à bord de la frégate *La Clorinde* pendant la campagne de 1886." *Revue maritime et coloniale* 93 (1887a): 398–430

– "42 photographies de Saint-Pierre, de Cap-Breton, de Terre-Neuve et du Labrador." 1887b

– "Un été le long de la côte française de l'île de Terre-Neuve." *Revue scientifique* 39 (1887c): 325–32

– "Sur la mesure de la densité des eaux de mer. Considérations générales sur le régime des courantes marins qui entourent l'île de Terre-Neuve." *Annales de chimie et de physique.* 6e série, 14 (1888a): 289–337

– "Observations sur le Gulf-Stream." *Comptes rendus de l'Académie des Sciences* 105 (1888b): 862

– "Carte bathymétrique et géologique des Bancs de Terre-Neuve, d'après les cartes de la Marine française Numéros 1437, 1839, 3437, 3855)." (Campagne de la frégate *la Clorinde* 1886). *Bulletin de la Société de Géographie.* 2e trimestre (1889a)

– "Considérations sur la structure et la génèse des bancs de Terre-Neuve." *Bulletin de la Société de géographie de Paris,* 7e série, no. 10 (1889b): 203–41

– *Un voyage à Terre-Neuve.* Paris and Nancy: Berger-Levrault 1891. 1 vol-ume in 8°

– "Les dépôts sous-marins." *Revue scientifique* 50, no. 4 (1892): 104–10

– "Les derniers volumes des Reports du *Challenger.*" *Annales de Géographie* 17 (15 juillet 1896): 500

– "Carte bathymétrique de l'archipel des Açores." *Comptes rendus de l'Académie des sciences* 128, no. 24 (1899): 1471–3

– "Projet d'une carte générale des profondeurs océaniques." *Bulletin trimestriel Société de géographie de l'Est (N.S.)* 22 (1901): 5–22

– *Atlas bathymétrique et lithologique des Côtes de France.* Montauban :

Association pour l'avancement des sciences 1902
– *Échantillons d'eaux et de fonds provenant des Campagnes de la Princesse-Alice*
(Résultats des campagnes scientifiques du Prince Albert). Facs. xxii,
76 pages, 3 planches, 1902
–Letter from Thoulet to Albert I of Monaco, 27 November 1913.
Archives du Musée océanographique de Monaco.
Tréfeu, Étienne. *Nos marins.* Paris: Berger-Levrault 1888
Tuck, James, and Ralph Pastore. "A Nice Place to Visit, but ... Prehis-
toric Human Extinctions on the Island of Newfoundland." *Canadian
Journal of Archaeology* 9, no. 1 (1985): 69–80
Turgeon, Laurier. "Colbert et la pêche française à Terre-Neuve." In
Roland Mousnier, dir. *Un nouveau Colbert: Actes du Colloque pour le tricen-
tenaire de la mort de Colbert.* Paris: Editions Sedes 1985: 255–68

Unger, Richard. "The Archaeology of Boats: Ships of the Vikings."
Archaeology 35, no. 3, (May–June 1982): 20–7

Vallaux, Camille. "Notice sur Julien-Olivier Thoulet (1843–1936)."
Bulletin de l'Institut Océanographique 702 (30 Juin 1936): 1–28
Vinner, Max. "Unnasigling – the Seaworthiness of the Merchant Vessel."
Viking Voyages to North America, ed. Birth L. Clausen, trans. Gillian Fel-
lows Jenson. Roskilde: Viking Ship Museum 1993: 95–108

Wallace, Stewart. *Macmillan Dictionary of Canadian Biography.* (4th edi-
ton) revised, enlarged, and updated by S. Wallace. Toronto: Macmillan of Canada 1978
Wilkshire, Michael, ed. and trans. *A Gentleman in the Outports: Gobineau
and Newfoundland.* Ottawa: Carleton University Press 1993

Index